GW00716610

CONSERVATION OF FISH AND SHELLFISH RESOURCES: MANAGING DIVERSITY

CONSERVATION OF FISH AND SHELLFISH RESOURCES: Managing Diversity

J.E. THORPE
Institute of Biomedical and Life Sciences, University of Glasgow, Glasgow, UK

G.A.E. GALL
Department of Animal Science, University of California, Davis, California, USA

J.E. LANNAN
Hatfield Marine Science Center, Oregon State University, Newport, Oregon, USA

and

C.E. NASH
Rolling Bay, Washington, USA

ACADEMIC PRESS
Harcourt Brace & Company, Publishers
London San Diego New York Boston
Sydney Tokyo Toronto

ACADEMIC PRESS LIMITED
24-28 Oval Road
LONDON NW1 7DX

U.S. Edition Published by
ACADEMIC PRESS INC.
San Diego, CA 92101

This book is printed on acid-free paper

A catalogue record for this book is available from the British Library

ISBN 0-12-690685-8

Printed and bound in Great Britain by Hartnolls Ltd, Bodmin, Cornwall

In memory of a wry humorist,
Giora Wohlfarth,
killed in a car accident in autumn 1994.

CONTENTS

PART TWO: Management examples

Contributors

B.E.Ballachey National Research Council, Board on Agriculture, 2101 Constitution Avenue, Washington DC 20418, USA

A.E.Eknath International Center for Living Aquatic Resources Management, MC P.O.Box 2631, Makati, Metro Manila 0718, Philippines

G.A.E.Gall University of California, Department of Animal Science, Davis CA 95616, USA

D.Hedgecock University of California, Bodega Marine Laboratory, P.O.Box 247, Bodega Bay CA 94923, USA

J.E.Lannan Oregon State University, Hatfield Marine Science Center, Newport OR 97365, USA

C.E.Nash P.O.Box 4606, Rolling Bay WA 98061, USA

L.Stradmeyer Wester Balrobbie, Killiecrankie, Perthshire PH16 5LJ, Scotland, UK

J.E.Thorpe Institute of Biomedical & Life Sciences, University of Glasgow, Glasgow G12 8QQ, Scotland, UK

G.W.Wohlfarth† Agricultural Research Organisation, Fish & Aquaculture Research Station, Dor, D.N. Hof Hacarmel 30820, Israel

FOREWORD

The needs of a rapidly expanding human population pose an immense challenge. By early in the 21st century the global population will exceed 6 billion people. Satisfying their food and nutritional needs will require an unprecedented and sustained agricultural output. It has been estimated that the total output in the decade following 2010 will be equal to that since the dawn of agriculture. This will require not only increased production, but the development of new, highly productive food resources. It is in this context that scientists and others have looked to the world's marine and freshwater environments as important sources of nutritional protein.

Fish and shellfish comprise annually nearly 70 million tons of the world's edible animal protein. They account for about one-third of the animal protein consumed in the developing world. However, over the past four decades, many individual stocks of fish and shellfish have been overexploited, almost to the point of extinction. These activities pose severe threats to the world's aquatic animal resources. Modern and increasingly efficient methods have enabled factory ships and floating processing plants to harvest massive quantities from a resource the extent and resilience of which is relatively unknown. Pollution from effluents carried by rivers or industrial wastes released into inland watersheds and estuaries, and from accidental or deliberate discharges at sea also threaten the organisms that inhabit the world's lakes, streams, rivers and oceans.

As vast as they are, the world's oceans and their resources no longer can be regarded as limitless or immune from the effects of human activities. All natural populations of fish and shellfish are affected by a variety of commercial and environmental pressures. It is not surprising, therefore, that interest in managing fisheries or individual stocks has grown, as exemplified in the resurgence of aquacultural practices and their extension to new species. However, only about 15% of the global fish and fisheries production comes from such culture. Natural populations still supply the bulk of aquatic animal protein. Even aquaculture is not without potential adverse consequences. Escape or accidental release of cultured populations can have significant effects on local populations either by altering natural gene pools or by introducing exotic species that compete with native species.

The management practices for natural populations are less intensive than those used in aquaculture, and depend more on setting and observing limits on factors such as catch, net size, fishing areas, and seasons. Their aim is to limit or prevent the harvest of smaller and less mature individuals. However, it is realized now that even these methods can result in selective pressures that may lead to changes in the genetic structure of populations.

Except for those species that are farmed at present, little serious effort has been expended to maintain and protect genetic stocks for replenishing areas that have been overfished or reduced severely by pollution or other factors. Further, some of the actions undertaken to maximise fish harvests threaten genetic diversity. Although vast, the world's aquatic resources are not readily visible in the way that a satellite image can track the loss of tropical forest. As a consequence there is much about these resources that remains unknown.

This book presents the major issues surrounding commercial fish production, preservation and conservation. The four management studies included illustrate the breadth of challenges to be addressed if these resources are to be managed wisely and sustainably. The aquatic environment also harbours a wealth of plant and microbial life that may suffer the effects of pollution, overharvesting, and poor management. While a specific study of these is beyond the scope of this book, management strategies for the world's aquatic genetic resources should include them.

The book grew out of the discussions of the Committee on Managing Global Genetic Resources of the United States Natonal Research Council (NRC). That committee produced a series of four reports in its Managing Global Genetic Resources series that focused on genetic resources of identified economic value: Forest Trees; Livestock; Agricultural Crops and Policy Issues; and The US National Plant Germplasm System. The Subcommittee on Animal Genetic Resources, at the suggestion of its NRC staff officer, Dr Brenda Ballachey, examined the issues related to the management and use of aquatic animal genetic resources. The analyses of a working group subsequently established under the direction of Dr James Lannan contributed to the deliberations of the committee, and led ultimately to the independent development of this book.

The book is in two parts. The first part deals with issues and challenges in managing fish genetic resources and makes some recommendations. It reviews the needs and constraints, the impact of human activities, the problems posed by different aquatic species, and the present status of aquatic genetic resources management. The second part presents four management studies: the Atlantic salmon, cupped oysters, the common carp and Chinese carps, and the Nile tilapia. These are intended to provide greater detail and reference to the comparison of species management developed in the first part.

We commend the authors of this work, and in particular Dr John Thorpe, for their persistence in completing their study. The difficulties and challenges are great. Aquatic populations are mobile and occupy an environment that is technically very challenging. There is competition among nations to garner these resources, which are regulated less than perfectly by international laws, conventions and agreements. Only well-planned, scientifically sound conservation, management and use will enable society to continue to draw benefits from the world's marine and freshwater environments. The pressure of the increasing global population makes achieving this essential to future generations.

Peter R.Day, Director, AgBiotech Center, Cook College, Rutgers University,
New Brunswick, New Jersey. USA.

Michael Strauss, Director, Global Change Program, American Association
for the Advancement of Science, Washington, DC. USA.

ACKNOWLEDGMENTS

The original discussions which have culminated in the production of this book were held under the auspices of the National Research Council (NRC) of the United States National Academy of Science, in Washington, DC, USA. We thank the members of the Committee on Managing Global Genetic Resources for originating these discussions and inviting us to participate in them; and the NRC for hosting meetings of the Aquatic Animals Group, and for arranging the groups of reviewers who contributed valuable comments on earlier versions of this text. In particular we thank Drs Peter Day and Michael Strauss for their personal support, and for maintaining enthusiasm for the project throughout.

We are grateful to Dr Gideon Hulata for providing the photographs of carp, Dr Lars-Petter Hansen and David Hay for the photographs of Atlantic salmon, and Ms P Davis and the Pacific Coast Oyster Growers Association for the photographs of Pacific oysters; and to Dr Donatella Crosetti for contributions to an earlier draft of Chapter 8.

SUMMARY

Water covers more than two-thirds of the earth's surface. Beneath a variety of marine and freshwater ecosystems is a wealth of biological diversity, much of which is yet to be discovered or fully described. The multicellular metazoan species of the world's aquatic environments number at least 150,000. Although the number of aquatic species is fewer than all terrestrial species, the number of phyla in the oceans represents a breadth of diversity that is far greater.

The fish and fisheries products derived from the world's oceans, lakes, and rivers make important contributions to human nutrition and health. They are also a vital part of the economies of many societies. In 1991, the global commercial fisheries' harvest, including that from aquaculture and culture-based fisheries, was approximately 97 million metric tons. Of this amount, 71% was intended for human consumption. Imports and exports of fish and shellfish products worldwide in 1991 were valued at $82.0 billion. This included 13 million metric tons of crustaceans and molluscs harvested worldwide in 1991, with a value of $29.7 billion as fresh, frozen, dried, or salted commodities.

The world's aquatic animals comprise a resource that fulfils many human needs today and provides a source of genetic diversity from which future needs can be met. However, a variety of human activities threaten the existence of aquatic animal species. The majority of fish and shellfish are harvested from populations that live in natural environments. Overharvesting in these areas coupled with pollution, environmental degradation, and loss of habitat raise the potential for significant depletion or loss of important aquatic resources. For the many aquatic species not of commercial importance, such threats can be even greater. The introduction of exotic fish into lakes and streams can produce new fisheries, but at the expense of creating drastic upheavals in the ecosystem that can lead to the loss of endemic populations and species.

'Variability of biological matter is the *sine qua non* for the ability of organisms to cope with the uncertainty of the environment' (Conrad, 1983). The genetic structure of a species is the repository of that variability, and its conservation is essential to maintain the genetic continuity of that species in the face of continual environmental change.

This book addresses the need for increased efforts so that the management practices used in global fishery activities can be based on sound genetic principles. It stresses the biological diversity of aquatic species and the unique constraints to genetic management. Too often, the long-term effects of management practices on population genetics have been overlooked. To ensure that future human needs will be met requires greater consideration for managing and conserving the genetic diversity of the world's aquatic animal resources.

BARRIERS TO MANAGING AQUATIC ANIMAL GENETIC RESOURCES

Several factors limit effective management or conservation of the genetic resources of aquatic animals, ranging from a lack of scientific information or technology to the need for appropriate programmes, institutions, or national and international policies.

The Lack of Information

Little is known about the genetics and evolutionary biology of most aquatic species. Aquatic animals are far more difficult to observe than terrestrial species, and genetic investigations on aquatic animals have not received adequate priority or funding. A comprehensive inventory of aquatic species and their genetic diversity does not exist. For some of the deepest portions of the oceans, the barrier to acquiring information is the direct result of technological difficulties. For the coastal zone, as for the tropical rain forests of the terrestrial environment, it is probable that more organisms exist there than have been described.

The vast diversity of aquatic species and the differing means by which they must be conserved confound the development of appropriate management strategies and programmes. Only a small fraction of the aquatic species are considered to be economically important. A comparatively large amount of information exists for these species. To illustrate the diversity of genetic structure of aquatic animal populations, and the differing approaches necessary for their conservation, four species with widely differing life-styles are discussed in Part Two of this book, namely: a migratory fish, Atlantic salmon (*Salmo salar*); a sedentary invertebrate, cupped oysters of the genus *Crassostrea*; a group of cultured non-migratory fishes, the common carp (*Cyprinus carpio*) and the Chinese carps (silver, *Hypophthalmichthys molitrix;* bighead, *Aristichthys nobilis;* grass, *Ctenopharyngodon idella;* black, *Mylopharyngodon piceus;* and mud carp, *Cirrhina molitorella*); and a fish with a broad range of intrageneric interactions, the Nile tilapia (*Oreochromis niloticus*). These are used to illustrate the diversity of challenges that confront numerous economically important aquatic species. There are many more species that could become important as sources of food, industrial, or pharmaceutical products or as indicators of environmental health.

Lack of Institutional Leadership

The responsibility and jurisdiction for conserving aquatic genetic resources is often poorly defined. In many nations, the management of aquatic resources and the regulation of human activities that affect them are not within the domain of any single institution or sector.

Lack of Conservation Focus

Genetic considerations have not been given a high priority in the management of aquatic resources. Relatively few programmes are designed for the express purpose of conservation. Many programmes have been enacted primarily for environmental management or species protection. Although these programmes may incidentally provide some degree of genetic resource conservation, most of them are concerned with population size and treat all individuals in the population as being genetically equivalent.

STRATEGIES FOR CONSERVATION AND MANAGEMENT

No single management strategy exists for minimizing or ameliorating the genetic risks associated with activities that affect aquatic species. Conservation management programmes must be considered on a species-by-species, case-by-case basis, but within a broad ecosystem context. At the very least, management plans must consider the

- nature of the resources;
- characteristic components of diversity of the species;
- genetic processes influencing diversity; and
- natural and human events or interventions that may place the genetic resources at risk.

If management programmes fail to conserve genetic resources, more extreme policy and management actions must be taken to prevent further genetic deterioration or extinction of species. For each species or population a determination of the most appropriate *in situ* or *ex situ* conservation must be made.

INSTITUTIONAL EFFORTS TO MANAGE AQUATIC GENETIC RESOURCES

Many of the international bodies involved in conserving aquatic species are international fisheries commissions, such as the European Inland Fisheries Advisory Commission, the Committee for Inland Fisheries of Africa, and the Indo-Pacific Fisheries Commission. Some of them were established through conventions or treaties fashioned to prevent the overharvesting of certain species. Although the commissions were not founded to conserve genetic resources, most of them now recognize the importance of genetic resources, and some have sponsored genetic research to provide information for resource management decisions. Given the mandate, international fisheries commissions could make important contributions to conserving the genetic resources of species harvested in international waters, as well as migratory species that inhabit the waters of two or more nations.

National policies and legislation that affect aquatic species are broad and varied. In a survey of scientists worldwide, few of the respondents indicated that their nations considered genetic conservation to be a separate resource management concern. Most of them stated that their nations had some form of regulation to prevent overharvesting of fishery resources and to control pollution or other activities that alter aquatic habitats. Many also indicated that their countries regulate the importation of exotic species.

RECOMMENDATIONS

The conservation of aquatic animal genetic resources will require coordinated actions on a global scale that include multilateral agreements and unilateral actions to develop and implement appropriate policies. These activities must be reinforced and enacted at national, regional, and local levels. Success necessitates greater public awareness of the importance of aquatic genetic resources and the need to conserve their genetic diversity. Major research, education, and training efforts will be needed to develop and apply technologies and to disseminate information to managers, policymakers, students, and the general public. Our principal recommendations for addressing these issues follow. Additional information and recommendations can be found in the main text.

Increasing the Effort

Management and conservation of the genetic resources of aquatic animals should encompass a greater number of species.

Conservation of aquatic animal genetic resources does not receive adequate attention in resource management programmes. Frequently the aim has been to address the size of populations and implicitly assume that all individuals are genetically equivalent. The problem is compounded by a lack of clarity about national responsibility or governmental jurisdiction over aquatic resources. A significant amount of information exists for only a relatively few species.

Only about 2000 of the known aquatic animal species are used in some form of managed system. Genetic conservation and management efforts exist for very few. Conservation of the genetic variation within unexploited species is necessary to preserve future options for development and use, and as a part of global efforts to conserve biological diversity.

Conservation efforts must be founded on a sound base of information and knowledge. Essential to increased efforts is the development of an inventory of the world's aquatic resources. This endeavour must be supplemented by efforts to understand the basic biology, genetics, demographics, and ecology of aquatic animal species. Without this greater level of information, management decisions may inadvertently place potentially important or fragile aquatic resources at risk.

Research is needed on the biology, genetics, ecology, management, and conservation of a wide range of aquatic animal species.

The variety of form and function in the world's aquatic environments may rival that of the tropical forests, yet information about most aquatic species is lacking. Funding and support for genetics in aquatic animal research have not received high priority. There should be more research to monitor and counter the consequences of human activities and interventions so that management practices do not lead to unanticipated consequences born of scarce information about the species involved. New and emerging biotechnologies promise a capacity for characterizing and studying the genetics of aquatic animal populations.

Techniques for storing long-term *ex situ* collections of ova, sperm, or embryos of most aquatic organisms do not exist. They must be developed and broadly applied, especially for preserving species threatened with extinction. These collections can be important resources for producing and improving captive stocks.

Considering Genetic and Biological Diversity

Maintenance of the genetic diversity of aquatic animal species should be considered when management and exploitation practices are developed.

Historically, resource management has become an issue only after concerns about overharvesting or resource competition have been raised by the capture fisheries, which tally the bulk of the world's aquatic animal harvest. Information about the genetic factors and selective pressures that influence natural populations is needed to develop methods that are not detrimental to resource survival. Production-related selection and breeding decision making must consider the potential for adverse effects to natural populations. Finally, much greater evaluation of the biological, ecological, and genetic impacts of species introductions and transfers is needed to anticipate the potential for adverse consequences.

Developing Practices to Suit Needs

The conservation of aquatic animal resources must address the unique characteristics of individual species and populations.

The variety of relationships between the components of diversity, the methods of management, and the potential for genetic depletion to different aquatic species resulting from environmental influences, such as toxic pollution, must be evaluated on a species-by-species basis. Management plans must be tailored to the biology, ecology, genetic diversity, and population structure of the species and to the potential interactions with human activities and interventions with the genetic processes that restrict diversity, namely selection and drift, on the genetic stability of that species.

Developing Leadership

National and global leadership is needed to coordinate activities to ensure the conservation of aquatic genetic resources.

Typically, conservation programmes are developed, implemented, and enforced at national, regional, and local levels. The effectiveness of national programmes often depends on international cooperation and coordination of local, national, and international activities.

Conserving Resources Through Education

Education and training programmes must be strengthened, and attention should be given to heightening public awareness of the importance of conserving aquatic genetic resources.

The implementation of effective policies to conserve aquatic genetic resources will require major educational efforts. However, the research and educational programmes of many nations generally do not recognize genetics as an important adjunct to aquatic sciences training. Professional education and technical training programmes must be strengthened to enable resource managers, scientists, and technicians to develop and implement genetically appropriate policies and practices. There should be greater public awareness of the importance of aquatic genetic resources and the need for conservation. While a degree of public concern exists for some charismatic aquatic species, such as whales, the case for and constraints to conserving many less visible species are almost unknown to the public.

PART ONE

1
THE NEED TO MANAGE FISH AND SHELLFISH GENETIC RESOURCES

The earth's inland, coastal, and marine waters contain a remarkable diversity of animal species. Were it not for terrestrial arthropods, the 150,000 or more aquatic animal species would account for more than 90% of all recorded species of animals with cells differentiated into tissues and organs (Metazoa) (May, 1988). Although many living aquatic species have undoubtedly not yet been described, aquatic species represent all major groups of metazoans, and some groups are comprised exclusively of marine or aquatic forms.

The variety of form and function found in aquatic environments clearly represents a wealth of biological diversity (Ray, 1988; Thorne-Miller & Catena, 1991). The undiscovered diversity of species in the world's oceans may rival that of the tropical forests (Ray, 1988). If the number of higher taxonomic categories, such as families, orders, or phyla, represented by species are counted, a much greater spectrum of diversity is found in aquatic environments. The number of phyla represented in the earth's oceans is about twice those with terrestrial representatives (Ray, 1988).

Aquatic resources, and more specifically fish and shellfish, are an important source of animal protein for human nutrition worldwide, forming a major portion of the diet in many cultures. The world commercial fisheries' harvest for 1991 was 97 million metric tons, with 71% intended for human consumption (Food & Agriculture Organization, 1993a,b). Of the total harvest, 84 million metric tons were from marine waters while 13 million metric tons came from inland waters (Food & Agriculture Organization, 1993a). These amounts include global aquaculture and culture-based fisheries production.

Fish and shellfish form an important component of international trade. In 1991, global imports and exports of fish and shellfish products were valued at $82 billion (Food & Agriculture Organization, 1993b). Global demographic trends indicate that the demand for fisheries' products will continue to expand. Recent findings that demonstrated the effect of animal fat consumption on human health have resulted in rising per capita consumption of fisheries' products in some regions. Other increases in demand coincide with improvements in the standard of living. Greater consumption heightens the potential for aquatic species to be vulnerable to depletion.

> Variability of biological matter is the *sine qua non* for the ability of organisms to cope with the uncertainty of the environment (Conrad, 1983).

The genetic structure of a species or population is the repository of that variability. Its conservation is essential to maintain the genetic continuity and adaptability of that species in the face of continual environmental change. Little is known about future environmental conditions nor about the adaptive values of particular genetic attributes (alleles and allelic combinations) under those conditions. Hence there is a need to conserve as much as possible of existing genetic variability, and there are no realistic alternatives to this goal (Ryman, 1991).

This book describes the nature of aquatic genetic resources and discusses their status and conservation. Of particular concern is the degree to which aquatic species are at risk of losing their genetic diversity and the activities which can accelerate or mitigate these losses.

Evidence suggests that aquatic species are being lost or are undergoing genetic change because of natural and human actions (Table 1.1). In part these losses and changes are not novel. Species have experienced evolutionary change, including extinction, since the Precambrian era. However, it is not the change itself but the acceleration of the rate of extinction associated with human activities that is of primary concern, because of its devastating effects on biological diversity (Office of Technology Assessment, 1987; Myers, 1988; Wolf, 1988; Wheeler & Sutcliffe, 1990; Thorne-Miller & Catena, 1991).

Although some aquatic animals are exploited in multi-species fisheries, most are managed as single species populations. This book focuses on the maintenance of genetic diversity within species, but recognizes that this must be considered in the context of conservation of the biological communities and ecosystems within which particular species flourish. Adaptation implies a response to changed conditions: hence habitat conservation is essential to prudent genetic conservation, to ensure that the rate of habitat change does not exceed the adaptive capacity of those genetic resources.

SOURCES OF LOSS

Major changes can occur in fish and shellfish populations as the result of natural events or human activity. Shifts in currents, temperature changes, river sedimentation, earthquakes, and volcanoes can alter aquatic environments and disrupt their biological systems. The effects can be extensive and rival or exceed those attributed to pollution and environmental degradation. For example, weakening of the atmospheric circulation and fading of the Southeast Trade Winds along the South American west coast is a natural event that has occurred several times over the past century. This change leads to an inflow of warm, highly saline water to the Pacific coast off South America, an event called El Niño, which has been associated with serious damage to anchovy and other fish populations over a wide area.

Table 1.1. Threatened freshwater fish species (from Andrews, 1989)

Region	Freshwater fish species Total number	Number threatened
North America (USA & Canada)[1]	*c.* 700	157
South America	*c.*2700	12
Africa	2000+	46 (+250)[2]
Oriental Region	*c.*1200	15
Sri Lanka	64	10

[1]According to Miller *et al.*, (1989), North America has also suffered the extinction of 3 genera, 27 species and 13 subspecies of freshwater or diadromous fishes over the last 100 years.
[2]The 250 species listed in brackets are the cichlids of Lake Victoria. The threat to them from the introduced Nile perch (*Lates niloticus*) is still a matter of controversy (see p.27, Box 1).

The modification or degradation of environments that results from human activities can place aquatic species at risk of losing genetic diversity or becoming extinct. Environmental effects from events such as industrial pollution or the construction of hydroelectric plants may alter the physical, chemical, or biotic features of aquatic habitats (Nehlsen *et al.*, 1991). Changes in biological communities are likely to accompany physical or chemical alterations. The declining environmental conditions for aquatic animals, resulting from many varied human activities, increases the potential to accelerate the loss of aquatic genetic resources.

Some activities, such as overharvesting of natural populations or destruction of habitats, have been occurring for many years. Serious depletion has already been observed in some species. Because of the interdependence of species within a community, changes in populations of a given species may place other species of a community at risk (Ehrlich & Ehrlich, 1981; Spencer *et al.*, 1991; Thorne-Miller & Catena, 1991).

Intentional or inadvertent transfers and introductions of aquatic species into new environments have occurred widely during the past century. Often the transfers or introductions have been related to management actions intended to benefit human societies. Although some transfers have resulted in economic benefits, others have had devastating effects on the genetic resources of other species.

CONSTRAINTS TO MANAGEMENT AND CONSERVATION

Several factors complicate effective management and conservation of aquatic genetic resources. These relate to the need for greater levels of information and scientific knowledge, insufficient attention to aquatic genetic diversity by resource managers, and the magnitude of the task.

Inadequate Knowledge Base

Substantial genetic information exists for a few intensively studied and managed species, such as the salmonids, tilapia, and carp. Because most aquatic species live in relatively inaccessible locations, data on their basic biology is difficult to obtain, and virtually nothing is known of their genetics or population biology.

The knowledge base for managing the genetic resources of aquatic species is inadequate. Although sufficient information is available for a few individual species, there is a paucity of even basic biological information for most of the rest. Genetic research on aquatic animals has not been considered a high priority, has not been funded adequately, and is difficult to execute. The result is that the impact of changing environmental conditions cannot be assessed adequately, even where biological knowledge about a species exists. This general lack of understanding of the genetics and evolutionary biology of aquatic species limits the effectiveness of current policies.

Insufficient Attention

Aquatic animal genetic resources have not received adequate consideration in resource management. A British symposium on *Biology and Conservation of Rare Fish* (Wheeler & Sutcliffe, 1990) found it necessary to resolve *inter alia* to "draw the attention of national and international bodies to the urgent need to give fishes a conservation status comparable to that already given to birds, mammals and other vertebrates" (Wheeler & Sutcliffe, 1990: 217). Subsequently this resolution was adopted also by the American Fisheries Society at their annual meeting (Anonymous, 1991a). Most resource management programmes emphasize the size of managed populations, making the implicit assumption that every individual in the population is genetically equivalent. The information collected by management programmes is of demographic interest, but does not contribute significant knowledge about genetic resources. Funding for research on the genetics of aquatic animals has not received high priority. In addition, training programmes for aquatic resource managers do not include enough study of genetics and evolutionary biology. Most university curricula in fisheries and wildlife management either do not require courses in resource genetics or have only token course offerings. Professional certification programmes for fisheries scientists typically do not specify competency in resource genetics. Consequently specialists in the genetics of aquatic species are scarce.

The responsibility and jurisdiction for conserving aquatic genetic resources are often unclear. The nature of aquatic resources and their use is highly complex. Frequently aquatic animals are common property resources. Even where they are privately owned, they may be linked inextricably to common property resources on which they depend for brood stock. Although most nations have developed organizations to administer their common property, overlapping jurisdictions often limit agencies' abilities to satisfy their mandates.

Magnitude of the Task

All aquatic animal species are potentially important reserves of genetic variation which can be essential for their adaptation and survival, and which can provide significant benefits to society. However, fewer than 2000 of the more than 150,000 known aquatic animal species are used in some form of managed systems. Global concern about the loss of genetic variation and biological diversity (Office of Technology Assessment, 1987; Wilson, 1988) should extend to all unmanaged species, as well as to those currently exploited or managed for the production of food for humans and animals. Species currently unexploited may one day be useful; indeed many aquatic species now managed as food resources have been used widely only within the last century. Conserving genetic variation in unexploited species preserves many future options. These species may become important food resources in the future. They are potential indicators of environmental quality. With the growth of techniques for gene transfer, they can serve as important reserves of genetic variation. Finally, they may affect the conservation of other species that are managed or exploited currently and with which they interact.

It is not possible to develop a uniform management and conservation strategy for all aquatic genetic resources. Conservation needs for an aquatic species depend, for example, on its biology, genetic diversity, and the human activities that influence its evolution. For example, management of species, such as the tunas, depends on natural reproduction of wild stocks, whereas management of species such as trout, salmon, and catfish includes domestication and artificial propagation. The requirements for managing and conserving the genetic resources of each of these may vary considerably.

RECOMMENDATIONS

Management and conservation of the genetic resources of aquatic animals should encompass a greater number of species.

There are significant voids in scientific knowledge about the genetics of most aquatic species. As a consequence, policies and programmes to conserve aquatic animal species must often be based on opinion rather than on scientifically sound information. Making conservation and management decisions in the absence of scientific information may inadvertently place aquatic resources at risk. More detailed information is required on the basic biology, genetics, demography, and ecology of aquatic species. The information base should be expanded to include a broader and more detailed view of the biological diversity of aquatic species. Understanding of the basic biology, genetics, demographics, and ecology of aquatic species should encompass this broader view better.

Research is needed on the biology, genetics, ecology, management, and conservation of a wide range of aquatic animal species.

The paucity of information about the genetics of aquatic as compared with terrestrial species shows that this area of study has not received appropriate priority or funding. Research is required to develop practical methods to determine and monitor the genetic properties of aquatic species, to evaluate the genetic effects of human interventions and management actions, in order to conserve the genetic resources of populations and species of fish and shellfish.

2
HUMAN INTERVENTIONS AND GENETIC RISKS TO FISH AND SHELLFISH

Human activities that can result in damage to the genetic resources of fish and shellfish include the modification of environments; harvesting or culturing of aquatic resources for food or other uses; and transfers, introductions, or enhancement programmes for aquatic species (Food & Agriculture Organization, 1981; Nelson & Soulé, 1987).

MODIFICATION OF ENVIRONMENTS

The genetic constitution of an individual is its blueprint for survival and development. That development proceeds in response to environmental opportunity. The genetic resources of a species have been shaped by natural selection in accordance with its environment. Its genetic diversity within and among individuals provides resilience to survive environmental change. The scale and rate of most non-catastrophic natural environmental change can be accommodated by the genetic resources of wild species. However, the scale and rate of man-made change is often beyond the genetic capacity of species to resist. Because of this close tailoring of genotype to habitat, modification or degradation of environments can place aquatic populations at extreme genetic risk and may result in extinction (Frankel, 1974; Ray, 1988). Environmental modifications may include alterations of physical, chemical, or biotic features of aquatic habitats.

The construction of dams and harbours, and the dredging or filling of waterways, are examples of highly visible physical alterations of aquatic environments. Other alterations are less visible. Siltation of lakes, streams, rivers, estuaries, and coastal areas that follows land-based activities often results in drastic alteration of aquatic habitats. Similarly, alteration or removal of riparian vegetation results in changes in plant and animal communities, thermal regimes, and exposure to sunlight (Zalewski, 1991). Withdrawal of water for irrigation, industrial, or domestic use can result in altered stream volume and current velocity (Peters, 1982). Changes in chemical and biotic factors are likely to accompany all such alterations to the physical features of aquatic environments (Li *et al.*, 1987). Physical alteration of aquatic environments can heighten competition for space or food resources or eliminate critical habitat. Loss of reproductive habitat, including spawning and nursery areas, is particularly devastating for aquatic animals.

Chemical modification of aquatic environments results primarily from pollutants, which are introduced directly or come from runoff or other indirect sources. Fertilizers and pesticides applied in agriculture and forestry, industrial discharges, and domestic wastes are all serious pollutants of aquatic environments. Airborne pollutants, most obviously from industrial activity, may also pollute aquatic systems.

Water pollution places aquatic animals at risk in at least three ways (Warren, 1971). First, the biological oxygen demand and chemical oxygen demand of organic pollutants can deplete supplies available to aquatic animals. Second, introduced chemicals may be acutely or chronically toxic for species in an aquatic community. Third, pollutants may lead to mutations that result in lethal or damaging genetic changes.

Physical and chemical changes culminate in new environmental conditions that can result in permanent alterations of biological communities (Karr, 1981; Li *et al.*, 1987). The balance of species may change, species not formerly abundant may thrive, or existing species or populations may be eliminated. The loss of species from a community may have profound implications for the survival of other species in that community (Ehrlich & Ehrlich, 1981; Soulé, 1986).

Loss of a species can occur if the new environmental conditions exceed that species' tolerance or capacity to adapt. Altered temperatures, acute concentrations of toxic substances, and depression of dissolved oxygen concentrations are examples of detrimental conditions that can accompany induced environmental changes. Although species may encounter such conditions at any stage in their life history, for many aquatic species their reproductive and early life history stages may be especially sensitive.

Even where environmental modification does not result in total loss of a species, the new conditions may place genetic resources at risk by introducing novel selection pressures that could change life history characteristics or result in significant genetic changes. Reductions in population size due to poor survival increase the likelihood that random genetic drift could lead to loss of genetic diversity. Mutagenic pollutants may further add to the risk of loss of populations or the genes they contain. Hence, even in the absence of detailed genetic information, prudent habitat conservation is a prerequisite for the successful genetic conservation of species *in situ*.

HARVESTING AND CULTURING

Fish and shellfish production differs from other contemporary animal production systems. Except for the management of game birds, agricultural production of livestock and poultry is almost exclusively based upon the husbandry of fully domesticated species. In fish and shellfish production, the transition from hunter-gatherer to farmer-rancher is not as advanced. Some aquatic species are being brought into cultivation (aquaculture); others are exploited under systems that combine cultivation and hunting (culture-based fisheries). However, humankind's relationship with aquatic animals is still chiefly as a predator (capture fisheries).

Capture Fisheries

Capture fishery is the harvest of aquatic animals from natural populations. The 3000 or more species that are harvested by the world's capture fisheries are at modest risk of losing genetic variation as a consequence of exploitation (Food & Agriculture Organization, 1981; Nelson & Soulé, 1987; Smith *et al.*, 1991; Pullin, 1994). However, they are at extreme risk of being overexploited, which would reduce the number of reproducing individuals in breeding populations. Reducing abundance of reproducing individuals results in increased rates of inbreeding and reduced genetic diversity due to genetic drift, and thus increases the potential for further loss (Gall, 1987). Such deleterious genetic effects would be mitigated if over-fishing were corrected at an early stage.

Selective fishing practices born of a need for economic efficiency of harvesting, or to comply with regulations imposed on the fisheries, also pose selective pressures that may present genetic risks (Nelson & Soulé, 1987; Law, 1991). The use of specific gear, decisions about resource allocations, and regulations governing when and where harvesting is permitted are examples of management and harvesting practices that can remove certain phenotypes selectively from exploited populations. These practices may result in long-term loss of genetic diversity, through directional change. An example of such directional change was quoted by Mathisen (1989) in herring (*Clupea harengus*). The date on which the fish first appear in their spawning grounds has been retarded progressively from September to February over the course of the past 90 years. One interpretation was that the original population consisted of many units peaking successively in spawning time from November to March, but that unrestricted fishing affected the first segments the greatest, and they were the first to go.

The risk of losing genetic variation through exploitation and selective fishing practices may increase if concurrently a species is experiencing loss or modification of its reproductive habitat. For example, the physical destruction of estuaries and coastal areas, which are the breeding grounds and nursery areas of many species, can hamper the capacity of a population to recover and further compound the effects of exploitation.

Effects of Modernization

Within the past 150 years the harvesting of aquatic animals by capture fisheries has evolved from a subsistence occupation by individuals to a highly mechanized industry involving many national and international cooperative ventures. The ever increasing efficiency of fishing technology, with modernized gear and new methods to detect fish, has led to rapid depletion of stocks and the closure of fisheries, such as those for California sardines and North Sea herring (Ahlstrom & Radovich, 1970; Gulland, 1971; Nicolson, 1979). In extreme cases, fisheries observers have reported the disappearance of entire assemblages, such as groups of cichlid species in areas of Lake Malawi (Turner, 1977). Investigators have documented classic patterns in the responses of fish populations to exploitation, including numerical changes in catch per unit of fishing effort (Christy & Scott, 1965), and physiological changes in growth rate, age at first maturity, and fecundity (Ricker, 1981). These findings are warnings that inadequate fishing regulations pose significant risks to genetic diversity through their effects on population size (increasing drift), and spectrum of developmental patterns (increasing selection).

Capture fisheries often harvest significant numbers of non-target fish species, thereby exposing them to risk of genetic depletion. Frequently, modern practices have exacerbated this problem. Highly visible, timely examples include the incidental catch of turtles in the shrimp fisheries of the Caribbean and Gulf Coast of North America (Thompson, 1988), and the incidental mortality of dolphins in net fisheries for tuna (Allen & Goldsmith, 1981; Coe *et al.*, 1984).

Management of Stocks

Fishery scientists perceive managed fish populations to be composed of more or less discrete subpopulations, termed stocks. Historically the term has been used to describe intraspecific assemblages of fish or isolated breeding populations. By definition, stocks consist of randomly interbreeding individuals whose genetic integrity persists whether they remain spatially or temporally isolated as a group, or whether they segregate only for breeding and otherwise mix freely with other members of their species (Kutkuhn, 1981). However, for most species there is still difficulty in defining the extent of breeding units (Stock Concept International Symposium, 1981).

The methods developed for regulating fish and shellfish populations for optimal or maximal sustainable yields have assumed tacitly that each individual in a population is genetically equivalent. Consequently the genetic implications of exploitation practices have been ignored, and only recently have fishery biologists become aware of the need to base management practices on genetic principles (Allendorf *et al.*, 1986). This omission was a result partly of the inaccessibility of continental-shelf and high-seas species of major economic importance, and partly to the difficulty of measuring the genetic diversity and structure of the exploited populations. However, this situation is changing. Biologists now recognize the varying degrees of genetic complexity in the structure of exploited wild populations, and management agencies are applying the stock concept in fisheries management to take such genetic diversity into account (Gall *et al.*, 1989; Shaklee *et al.*, 1991).

An increasing trend among fishery managers is to use indicators of population structure to identify stocks as management units (Utter, 1991). The literature is replete with reports of quantitative and qualitative genetic variation in fish. Researchers have determined phenotypic and genotypic variances and correlations for quantitative traits in a few aquatic species (Gjedrem, 1983; Tave, 1986). Many reports cite allelic variation, revealed by protein electrophoresis, for numerous fish species (Ryman & Utter, 1986; Ferguson & Thorpe, 1991). Genetic factors can influence life history patterns, for example, the seasonal timing of reproduction (Ricker, 1972). Collectively, the evidence of genetic variation in fish populations provides a general framework for identifying differences in population structure among local populations within geographic regions.

Allelic variation revealed by electrophoresis has become the favoured method of surveying genetic variation in fish populations for stock identification. Protein electrophoresis has been called the most useful procedure yet devised for revealing genetic variation (Hartl & Clark, 1989), and the most fiscally prudent approach to population-level problems, such as stock assessment in fisheries (Buth, 1990). The methodology permits rapid and efficient observation of large numbers of individuals within populations, and in 50 or more loci within individuals. The amount of information that can be collected with a given amount of effort using electrophoresis is substantially greater than with alternative approaches to observing genetic variation (Utter *et al.*, 1986). More recent molecular genetic techniques that directly assess variation in the genetic material, such as restriction fragment length polymorphisms, although more expensive and time consuming, may yield significantly greater amounts of information (Allendorf *et al.*, 1986). These methods are likely to become more widely used in stock assessment, monitoring, and management programmes in the future (Ferguson & Thorpe, 1991).

The sensitivity of electrophoretic methods for determining if populations are organized into breeding units depends on the proportion of observed loci that are polymorphic. The proportion of polymorphic loci observed in electrophoretic studies is highly variable among fish species, including species that are closely related. Electrophoretic studies have demonstrated conclusively that some species (e.g., some Pacific salmons, *Oncorhynchus* spp., which have a relatively high proportion of polymorphic loci) are clearly organized into breeding units (stocks) that remain spatially or temporally discrete (Utter *et al.*, 1980; Utter, 1991). In other species, such as the skipjack tuna,

Katsuwonus pelamis, that show very little allelic variation, electrophoretic studies have not revealed discrete breeding units (Graves *et al.*, 1984). Based on electrophoretic and other biochemical evidence, such species may form panmictic breeding stocks, perhaps of worldwide distribution.

Although allelic variation is a useful indicator of breeding structure in some cases, breeding organization is indicated only if differences in gene frequencies are observed. The conclusion that breeding organization is absent does not follow in cases where gene frequencies do not differ among populations. One cannot exclude the possibility that more intensive sampling or observation of additional loci would reveal differences in allele frequencies.

Further, the lack of observable allelic variation in a species does not necessarily mean that other traits do not exhibit genetic variation, including quantitative traits. Unfortunately in most cases, little correspondence is apparent between patterns of allelic variation and continuously distributed quantitative traits, including life history traits (Lewontin, 1984; but see also Jordan & Youngson, 1991; Powers *et al.*, 1991). Therefore differences in breeding structures of populations as revealed by electrophoresis are not necessarily adaptive, that is, the differences may not be related to the population's capacity to survive and reproduce (Gould & Lewontin, 1979). A plausible alternative hypothesis is that the observed breeding organization results from founder effects and limited gene flow among populations. Thus, although electrophoresis provides a powerful tool for stock identification, it must be accompanied by studies of morphological, morphometric, and life history variation to provide more complete information on the genetic structure of populations.

Mixed Stock Fisheries

An especially confounding management problem is encountered with species such as Atlantic and Pacific salmon that are organized into discrete stocks. Stocks of these species are often heavily exploited while aggregated in common oceanic feeding areas. Harvesting of these aggregations (or mixed stock fisheries) may result in overexploitation of the numerically smaller stocks or stocks from less productive habitats, even if the more abundant stocks remain above a minimum viable population size. Mixed stock fisheries impose a high risk of depleting genetic resources of the less abundant stocks, which may lead ultimately to their extinction (Paulik *et al.*, 1967; Ricker, 1973; Hilborn, 1985). A similar management problem arises with mixed species aggregations of different fishes, such as flatfishes (Pleuronectidae), rockfishes (*Sebastes*), and cichlids (Cichlidae). One obvious management solution to the mixed stock fishery problem is to defer harvesting until the stocks have segregated. However, this type of regulatory policy is often unacceptable because it shortens the harvesting season, it may deny some harvesters access to the resource, and the quality (and thus the value) of some species declines as the fish approach sexual maturity.

Following Grant *et al.,* (1980), the Genetic Stock Identification (GSI) system has become a vital basic tool for the analysis of mixed stock fisheries, especially among salmonids (see for example, Gall *et al.,* [1989] and Waples *et al.* [1990]). Because genotypic frequency data differ between separate stocks, the relative contributions of component stocks in the mixture can be estimated probabilistically from protein electrophoretic samples derived from the fished population. The fundamental importance of such information for the rational genetic management of Pacific salmon populations has led to the collection of baseline data on stocks from throughout their geographic breeding range. Consequently, unlike almost all other aquatic animals, the body of genetic data on the *Oncorhynchus* spp. is now vast, probably only exceeded by that for *Drosophila* and humans (Waples *et al.*, 1990)

Aquaculture

For centuries people have maintained stocks of aquatic animals in captivity, particularly fish which have been used for recreation, decoration, and consumption (Hickling, 1962). Until about one hundred years ago, activity was largely in Asia and was limited to a relatively small number of freshwater species, especially carp. The practice of aquaculture spread to eastern Europe in the Middle Ages. The Romans were probably the first to impound fish and shellfish and to maintain marine species in coastal lagoons.

Growth of Aquaculture

Toward the end of the 19th century aquaculture changed with advances in technology, and people started to use it to propagate fish to compensate for habitat loss and overexploitation of natural stocks of some species (Bardach *et al.*, 1972). Aquacultural technology advanced concurrently with the expansion of mechanized fishing fleets and hydroelectric power generation. More recently aquaculturists have adopted highly advanced technology, paralleling developments in modern animal husbandry. Intensive systems of fish and shellfish production for direct consumption are now operating in many areas of the world.

Aquaculture production has increased steadily in recent years, and this trend is expected to continue into the next century. Aquaculture probably accounts for about 15% of present world fisheries production, up from 8.3% in 1984 (FAO, in press). In 1992 the world animal production in aquaculture was 13.9 million metric tons, valued at US$27.6 billion, up from 6.9 million metric tons and US$9.5 billion in

1984. Of this production in 1992, 84% arose in developing countries in Asia, China alone accounting for 50%, and India 10%. By volume and value respectively, fish accounted for 68 and 63%, molluscs 25 and 13% and crustaceans 7 and 24%. About 85% of all cultured finfish are non-carnivorous, being dominated by the carps. Of the carnivores, salmonids (44%), catfishes (26%) and yellowtails (10%) predominate. Among molluscs, oysters remained the most valuable harvest, but the volume of mussels produced was greater. Marine shrimp culture expanded by 325% between 1984 and 1992, particularly in Asia and Latin America, developing in super-intensive systems, and engendering major foreign exchange earnings as well as environmental and social disputes, water quality and disease problems, and crashes in production in some countries (FAO, in press). Based on projection of the FAO statistics, the present annual increase in aquaculture production is approximately 5% per year, while the yield from capture fisheries is increasing at no more than 1.5% per year.

Many freshwater, marine, and anadromous species are now cultivated for a variety of purposes. In all cases, aquacultural development poses potential genetic resource management risks, particularly as cultivation concentrates more and more on domesticated stocks. The risks are evident in two ways. First, the loss of natural populations poses problems of maintaining genetic diversity and sources of new genetic material for future aquacultural development. Second, the development of domestic stocks, with their inherent likelihood of escape into the natural habitat, poses some as yet unquantified risk to the genetic integrity of wild stocks (Hansen *et al.*, 1991). Domestication of captive stocks, including selection for production-related traits, can reduce the domesticated stocks' capacity for survival and reproduction in the natural habitat (Doyle, 1983). The perceived threat of genetic loss in wild stocks is four-fold: by breakdown of adapted gene complexes through interbreeding, by loss of native populations through competitive displacement, by loss of those populations through disease introduction, and by homogenization of population structure through swamping a region with a common gene-pool (Hindar *et al.*, 1991). A similar problem is posed by the deliberate releases of hatchery fish for augmentation of wild stocks, unless care has been taken to ensure the equivalent diversity and genetic compatibility of the released and recipient stocks (see below).

Genetic Conservation in Aquaculture

Aquaculture is in a unique position with regard to the need for genetic conservation programmes and the risks involved in ignoring the genetic consequences of aquacultural development. Generally, most species under cultivation have been sampled only recently (within the last 30 years) from the wild, so the wild stock is the source of genetic material.

Most species under cultivation are also harvested in capture fisheries. Thus, there are political and social concerns regarding access to the genetic resource as well as arguments about the genetic impact of released domesticated fish on the genetic integrity (fitness) of wild stocks (NASCO, 1990; Waples *et al.*, 1990; Hansen *et al.*, 1991). Finally, there is the unique system of using aquacultural production to compensate for losses in natural production or to enhance the natural productivity of capture fisheries. Aquaculture provides opportunities to reduce exploitation and thus conserve wild stocks on the one hand, while possibly exposing them to some level of risk on the other. The risks relate to the different selection pressures in the hatchery and natural environments.

Other genetic risks are inherent in managing captive populations. Often, brood stocks are derived from narrow samples of the ancestral stock, either through limited initial sampling or because of low reproductive success of captured fish. Biologists generally expect that high levels of genetic drift and inbreeding will accompany domestication and stock maintenance for many species. For economic reasons most brood stocks are maintained with limited effective population sizes, and generally fear of the spread of infectious diseases has prevented the exchange of genetic material among producers. Taken together, founder effects, genetic drift, intentional selection, and inadvertent selection during culture are likely to reduce the genetic diversity of a stock markedly.

The management of risks associated with aquacultural production, either for direct consumption or for culture-based fisheries, is not well developed. One of the greatest risks is loss of the wild populations that supplied the original germplasm of captive brood stocks. Because current captive brood stocks are unlikely to possess a diverse genetic sampling of wild gene pools, correcting future genetic degradation of captive brood stocks will depend on resampling wild sources. However, if the wild stocks have experienced losses of genetic diversity due to overexploitation, loss of habitat, or other causes, then the reservoir of ancestral material will be lost forever.

Managers of natural resources and aquacultural producers face formidable constraints in establishing and maintaining reserves of genetic diversity. First, aquacultural genetics is a young discipline. Trained aquacultural geneticists are in short supply. While plant and animal breeders have developed a tradition of genetic management and work routinely with domesticated species, aquacultural breeders are for the most part working with animals only a few generations removed from their wild ancestors.

Second, substantial capital and operating costs are associated with brood-stock maintenance. The specialized facilities (such as the Atlantic Salmon Broodstock Centre at Kjerksæterøra, Norway) and skilled labour required to maintain a genetically diverse brood stock, are beyond the financial means of many producers. Because most cultured species are highly fecund, some producers may think they can minimize the size of brood stocks. This is a misconception that would lead to large numbers of progeny being derived from only a few parents, and thus of restricted genetic diversity. The result can be an unexpected genetic divergence of aquaculture stocks from their ancestral populations, as probably occurred in the hatchery stocks of coho salmon (*Oncorhynchus kisutch*) studied by Simon *et al.* (1986).

Finally, and perhaps most critically, confusion exists regarding who is responsible for and has jurisdiction over maintaining genetic diversity. Until adequate germplasm protection and collection programmes have been established, wild populations must continue to serve as the principal reserves of genetic material. Thus conservation of genetic resources is not the exclusive domain of either aquaculture or natural resource managers.

Models for Conservation in Aquaculture

There are examples of established production brood stock programmes that can serve as models for the management of fish genetic resources. A government-sponsored programme in Hungary propagates 18 landraces of common carp, *Cyprinus carpio*, continuously (Bakos, 1976). The programme maintains nine domestic landraces and nine exotic landraces imported from Europe and Asia. The breeding centre provides fingerlings for aquaculture, for stocking impoundments for recreational fishing, and for restoring and enhancing depleted natural populations. The programme evaluates the performance of all landraces and their hybrid combinations continuously.

Another example of an established brood-stock programme is in Norway. Within the last 30 years, success with hatchery practices for rearing large quantities of Pacific salmon (*Oncorhynchus* spp.) and Atlantic salmon (*Salmo salar*) has led to the development of intensive farming in marine waters. This industry began in northern Europe (Norway and Scotland), and has since been taken up at favourable sites elsewhere in the temperate regions of both the northern and southern hemispheres. Already the annual commercial production of Atlantic salmon in aquaculture is more than 20 times the maximum recorded worldwide harvest of the species, and production in the industry is still increasing every year.

In Norway, the annual production of farmed Atlantic salmon was approximately 60,000 metric tons in 1986, reached 115,000 tons in 1989 (Bergan *et al.*, 1991), and approximately 200,000 tons in 1993. An industry-sponsored programme maintains a large production brood stock developed from wild progenitors collected from several riverine systems. Because of the relatively high fecundity of Atlantic salmon, the brood stock provides a surplus of progeny beyond that needed to maintain genetic diversity. The surplus juveniles are sold to commercial producers, and the revenues are used to amortize the investment in facilities, to pay the operating costs of the breeding programme, and to maintain the brood stock (T.Refstie, Institute of Aquaculture Research, Sunndalsøra, Norway, pers. comm., 1988).

Culture-Based Fisheries

The harvesting of natural and captive populations are combined in culture-based fisheries. Captive populations of young fish (such as the Pacific salmon, carp, tilapia) and shellfish (oysters, clams, abalone, lobster) are reared in hatcheries and released into open waters to supplement the natural populations and to be taken in the capture fisheries. Conversely, natural populations of young fish (such as milkfish and mullet) and shellfish (shrimp, prawn) are harvested and stocked in ponds for grow-out. Devices such as artificial reefs and rafts are used to attract and keep fish, or to provide surfaces for sessile shellfish, such as oysters and mussels, from which they are subsequently harvested. Such practices produce the culture-based fisheries that, particularly in the case of shellfish, constitute much of the total harvest of these species.

Culture-based fisheries have been developed frequently for species where natural production cannot satisfy the demand for commercial or recreational harvesting. For example, the intense fishing pressure on Pacific and Atlantic salmon species, combined with human activities that reduce the available reproductive habitat or impede migration, has led to the development of large-scale hatchery programmes that release fish mainly to support capture fisheries (Thorpe, 1980). Hydroelectric development on major river systems in North America and on major tributary rivers that enter the Baltic Sea have resulted in extensive losses of spawning and nursery habitat for salmonids (Nehlsen *et al.*, 1991). Salmon fisheries in these areas are now highly dependent on hatchery programmes that were implemented to mitigate the alteration or loss of habitat. In northern Japan and Russia, the growth in volume of salmon fisheries depended on artificial propagation to ensure increased availability of fish to the traditional high seas, near-shore, and terminal fisheries. Many culture-based fisheries began as limited duration programmes to restore depleted resources. Continuously growing demands resulted in their expansion beyond the original plans. Once aquaculture practices are established to supplement fisheries they tend to endure.

The genetic risks inherent in both capture fisheries and aquaculture are also present in culture-based fisheries. The mixed-stock fishery problem described earlier is also noteworthy here. Some culture-based fisheries, such as many of those for Pacific and Atlantic salmon, harvest aggregations of naturally and artificially propagated fish. These fisheries risk overharvesting the naturally reproducing stocks if the hatchery stocks are more productive than the wild populations. To preclude the loss or even extinction of the genetic resources of wild populations harvested in mixed fisheries consisting of both hatchery and wild fish, harvest allocations must be based on the reproductive and growth rates of the less productive, usually the wild, stocks.

Culture-based fisheries also present unusual concerns that result from potential intraspecific and interspecific interactions of artificially propagated fish with indigenous species. In this book, they are considered consequences of introduction.

INTRODUCTION OF SPECIES

Another human intervention that can influence genetic variation greatly is the introduction or transfer of exotic species or varieties to new geographic environments. This has been a common practice worldwide for almost 150 years, and is frequently practised in culture-based fisheries and enhancement programmes. Three types of introductions or transfers may be involved: (1) an exotic species, (2) an exotic stock of an indigenous species, and (3) a hatchery stock developed from an indigenous population that may be considered an exotic stock if it has become differentiated genetically from the indigenous stock.

Exotic Species

Many thousands of exotic species have been transferred or introduced. In some cases the results have been successful, if measured in economic terms alone. Examples are the establishment of aquaculture industries for shrimp and prawn in countries where the species are not indigenous, and the creation of recreational and capture fisheries in certain rivers and inland lakes. Other introductions were economically unsuccessful, such as the attempted introduction of Pacific salmon, *Oncorhynchus* spp., in various countries in South America (Joyner, 1980), or the transfers of juvenile flatfish and cod from breeding grounds to new coastal areas at the turn of the century (reviewed by Shelbourne, 1971).

Many transfers and introductions have been intentional, usually to establish or enhance commercial or recreational fishing. Sometimes, unintentional introductions have companied human manipulations of aquatic ecosystems or other activities (for example, organisms that attach to the bottom of ships). The colonization of the parasitic marine lamprey, *Petromyzon marinus*, in the Great Lakes of North America following the opening of the Saint Lawrence Seaway, is a well-documented example that had devastating consequences for indigenous fish populations (Smith, 1971). More recently, invasion of the zebra mussel has begun to create difficulties in that ecosystem (Roberts, 1990).

Introduction of predatory Nile perch (*Lates* spp.) into Lake Victoria in 1958 (Barel *et al.*, 1985) is an example of a thriving introduced fish that had rapid and devastating consequences for indigenous species. Predation by the Nile perch has led to depletion of the indigenous cichlid populations, to the point that some species no longer appear in the traditional fisheries. It has been estimated that more than 200 cichlid species may be in
Box 1.

The Introduction of Nile perch to Lake Victoria.

Lake Victoria in Africa, bordered by Tanzania, Uganda and Kenya, is the largest body of fresh water in the tropics. Fishing has always been important for people living in the area, both as a form of income and as a source of food. When the catch of tilapia and other endemic species declined during the 1950s, efforts were made to stock the lake with other fish in order to increase production. Starting in 1958, the Nile perch (*Lates niloticus*) was introduced into the lake. This was a reintroduction, as the species was abundant there and in Uganda's Lake Edward about 25 million years ago. One result, some scientists have claimed, has been the depletion of indigenous haplochromine cichlid species, to the point where some no longer appear in traditional fisheries. More than 200 cichlid species are estimated to be in danger of extinction, threatened by the predatory nature of the Nile perch.

But the issue may not be as clear-cut a case of wrongful introduction of an exotic species as it appears. At the same time that the perch was introduced, changes in fishing materials, pressures for higher catches from growing human populations in the area, and misinformation about maximum sustainable yields conspired to reduce the population of endemic fish species in the lake. In the 1950s, terylene and nylon gill nets were introduced: they cost more to start with, but were more efficient and lasted longer than the flax ones that had been used. Fishermen started to use outboard motors on their boats, to extend their range and compensate for declining yields in areas close to the shore. Although earlier the governments of Uganda and Tanzania had restricted mesh size to 127 millimetres to avoid catching immature endemic tilapia *(Oreochromis esculentus)*, the regulations were difficult to enforce, and were repealed in 1957.

The results of these changes in technologies and fishing practices was a serious depletion of the stocks of tilapia and other endemic species, as smaller fish and broodstocks were being caught. In the early 1960s gill nets with mesh sizes less than 90 millimetres were being used, and by late in that decade the size was down to 38 and 46 millimetres, to harvest smaller fish than had been exploited previously. The spreading use of beach seine nets, which harvest mature and juvenile tilapia and juvenile Nile perch, also depleted the stocks. With the decline of *O.esculentus* in the early 1970s, the numbers of the large predator fish such as *Bagrus* spp., and *Clarias* spp., increased, long before the introduced Nile perch was fully established.

By the early 1980s, cast nets below 76 millimetre mesh were being used, and were catching populations of mature Nile tilapia and juvenile Nile perch. Furthermore,these nets interfere with the breeding activities of both species. Trawl nets commercially since the 1970s have destroyed the population of indigenous cichlids in the Kenyan and Tanzanian parts of the lake, and medium gill nets used at the mouths of rivers have removed fish vital to the species' health at the beginning of their spawning migration. The misinformation about maximum sustainable yield stemmed from a visit in 1957 by an expert from an international organization, who estimated erroneously that the potential yield of *O.esculentus* and *O.variabilis* was 1,800 tons. This led to the expenditure of considerable sums to expand the market at Jinja, Uganda, despite declining catches. The introduction of Nile perch and tilapia at least saved the money spent on this market from being wasted, and helped sustain the livelihoods of fishermen in that area for a while.

One advantage of the introduction of these exotic species that is often lost in the controversy about their impact is the acceptability of the fish to the local population. Nile perch can be fried in its own fat, saving the expense of cooking oil. This and the Nile tilapia are large fish, compared with the small bony haplochromines, and hence are more popular with the consumers.

A full assessment of the effects of adding exotic species to endemic fish populations in lakes must consider both other pressures on existing stocks, and the impact on the livelihoods and food supplies of local people, so as not to draw hasty conclusions about the introduced species. In this case, the verdict about the results of introducing the Nile perch is a mixed one.

[Material from Acere (1988).]

imminent danger of extinction because of this single introduction (Barel *et al.*, 1985; Ribbink, 1987). (See Box 1).

The International Council for the Exploration of the Sea (1984) has proposed a code of practice for regulating exotic introductions of aquatic species. Most nations now have some form of policy, although levels of implementation and enforcement vary (Welcomme, 1988). Most policies agree that :

- governments should regulate exotic introductions stringently;
- conservative management is required;
- conservative management dictates that intentional introductions be allowed only after exhaustive analysis of both the ecological and genetic risks; and
- governments must make every effort to prevent unauthorized, accidental, and inadvertent introductions.

Exotic Stocks of Indigenous Species

The concerns about genetic interactions between introductions of exotic stocks and indigenous populations of the same species follow from the assumption that local populations are optimally adapted. If exotics interbreed with indigenous populations, an introgression of nonadaptive genes from the former into the latter may occur, resulting in hybrid progeny that may be less well adapted for survival and reproduction. The ultimate concern is if the hybrid stock would reduce genetic diversity relative to the indigenous stock (Hansen *et al.*, 1991; Cloud & Thorgaard, 1993), as has been shown to occur among some salmonids (Allendorf & Leary, 1988). As Ryman (1991) has pointed out, a tension exists between the need for protection of natural fish populations and the rapidly expanding levels of aquaculture causing increased migration into those populations. He has suggested (Ryman, 1991) that levels of introgression acceptable in a conservation context should be related to those occurring spontaneously, through natural migration between populations, which can be estimated from the fixation index F_{ST}, a measure of the variation of gene frequencies at selectively neutral loci, approximated by:

$$F_{ST} = 1/(4Nm + 1)$$

where N = population size, m = the migration rate, and Nm = the number of migrants per generation. It is the number of migrants rather than the rate which is critical, because the effect is the result of the balance between migration and drift. F_{ST} can be estimated from gene frequency differences (Ryman, 1983; Waples, 1987), and the corresponding value of the number of migrants per generation (Nm) used as a first attempt at assessing the level of introgression acceptable.

Hatchery Stocks

In cases where the introduced stock is a hatchery stock that has become genetically differentiated from the indigenous population from which it was developed, the differentiation results from hatchery management practices and not from any inherent properties of hatcheries. There is much scope for the improvement of hatchery management, especially in conservation hatcheries, to ensure that the initial genetic diversity is comparable to that in the wild stock to be augmented, and that the products of protected rearing are ecologically competent when released (Huntingford & Thorpe, 1992; and chapters in Thorpe & Huntingford, 1992). This implies goals different from hatcheries producing fish directly for human consumption, where domestication is desirable. As noted earlier, often hatchery brood stocks are founded with limited numbers of individuals of unknown parentage. Therefore a hatchery stock may begin to diverge genetically from its wild progenitors at the time of drawing the founding stock. Genetic drift and inbreeding may continue to accumulate in subsequent generations unless hatchery managers take specific preventive measures. Unless they develop a pedigree for the hatchery brood stock, disproportionate representation of some lineages and elimination of others is highly probable.

Intentional and inadvertent brood-stock selection may result in further divergence. For example, the timing of ovulation in salmonids has a genetic component. A frequent selective practice in salmonid hatcheries employed in culture-based fisheries is to use gametes from early-spawning fish to ensure filling the hatchery quota. The response to this selection is an earlier spawning date and reduced duration of the spawning season.

This discussion reveals that interpretations of the empirical evidence are largely speculative. A more rigorous evaluation would be possible only if pedigrees existed for hatchery stocks. Until hatchery managers understand the necessity of establishing pedigrees, the genetic management of enhancement programmes and culture-based fisheries will continue to be based on conjecture.

The use of hatcheries in culture-based fisheries presents managers with a real dilemma. Interactions of hatchery and wild fish may expose the latter to some level of genetic risk, but the use of hatcheries may be the only tractable approach to genetic conservation, including the genetic rehabilitation of depleted stocks, conservation of stocks exposed to permanent and severe alterations of habitat, and conservation of stocks harvested in mixed stock or mixed species fisheries. When a fraction of a depleted or environmentally threatened wild stock is reproduced artificially in a hatchery, to support that stock through protected rearing and release of the offspring into the wild, even though no exogenous genes are involved, there are still genetic hazards through changes in the total effective number (N_e) of the breeding population. Ryman (1991) and Ryman & Laikre (1991) have shown that this number is given by the following expression:

$$1/N_e = x^2/N_c + (1-x)^2/N_w$$

where x is the offspring from the captive partners, and (1-x) is that from the wild ones, and N_C and N_W are the effective numbers of captive and wild parents respectively. They pointed out that the total number is only equal to the sum of the effective numbers of wild and captive parents if the fraction of the artificially produced progeny is:

$$N_c/(N_c + N_w)$$

More seriously, this implies that supportive breeding may reduce the effective number far below what it would have been without any such support. Also, It should be noted that the calculation of effective numbers of wild parents is complicated in the case of species such as salmonids with overlapping generations (see Waples *et al.*, 1990; Waples, 1991).

When the cultured and wild stocks differ, managers of culture-based fisheries have two options for ameliorating the risks of interactions of hatchery and wild fish. One is to isolate the reproductive stages of the stocks, thereby preventing direct genetic (but not ecological) interaction. This form of isolation is possible with some, but not all, aquatic species. The second option is to merge the two stocks by introducing a representative sample of the wild fish into the hatchery brood stock in each generation. In practice, neither of these options is used widely; stocks are usually graded for availability and hatchery performance rather than to minimize genetic risk. However, whatever methods are used, it is vital to characterize genetically the hatchery and potentially interacting stocks before any mixtures can occur, and then to sample them regularly thereafter to monitor genetic changes. The percentage of heterozygous individuals (*H*), averaged over all measurable loci, is a convenient measure of genetic variability in a population. Management should aim to maintain the level of genetic variability in the wild population.

RECOMMENDATIONS

The conservation of genetic diversity must be an integral part of policies and programmes that affect aquatic animal resource development.

National policies that affect fisheries and aquaculture development can be extremely broad and varied, and may originate at all levels of government. They may be enacted in response to concerns about production, environmental management, or species protection, and are related predominantly to economic or ecological rather than genetic conservation.

Policies, such as those regulating harvests in capture fisheries, must be sensitive to the potential for adverse unintended consequences to peripheral species. Genetic conservation must be given greater weight in formulation of fisheries and aquaculture policies.

Culturing of captive stocks can have profound effects on the genetic diversity of aquatic animals. Where selection and breeding have been employed to develop populations more suited to production conditions, the potential for these to affect natural populations adversely must be addressed. Clearly this will require better understanding of the genetic diversity and population structure of wild and captive populations.

Maintenance of the genetic diversity of aquatic animal species should be considered when management and exploitation practices are developed.

The historical pattern in capture fisheries management is that typically initiation of harvesting has preceded management. The harvesters rather than the resource managers discover and develop fisheries. Management is implemented only in response to societal concerns about overharvesting, competition for common property resources, or both.

A contemporary management viewpoint is that fisheries stocks are subpopulations that have become genetically distinct and are adapted optimally for survival and reproduction in their environments. The assumption that differences in patterns of genetic variation between stocks indicate adaptations to different environments thus engenders a conservative approach to fisheries management that reduces the risk of inadvertent loss of adaptive genetic variation.

Fisheries management is in urgent need of methodologies and information that delineate clearly the genetic factors that influence variations in life history patterns. A better understanding of how information from genetic stock identification (GSI) and molecular methods relates to observable variations in morphology, ecology, and life history is needed.

The introduction and transfer of aquatic species should be regulated strictly by governments and should not be permitted without careful analysis of the potential ecological, biological, and genetic risks.

Often, serious environmental effects have accompanied transfers and introductions, whatever the economic result. However, the evaluation of the genetic and ecological impacts of exotic introductions on other genetic resources is often inadequate and uncertain. Even under the best of circumstances, unforeseen problems may arise. The introductions of common carp and brown trout (*Salmo trutta*) to North America and of rainbow trout (*Oncorhynchus mykiss*) to Europe and South America are examples of introduced species that have colonized successfully. Unfortunately the degree to which the colonization was at the expense of other fish species can never be documented fully. Other introductions, such as that of predatory Nile perch (*Lates* spp.) into Lake Victoria, have been devastating to native species and have altered traditional fisheries.

3
THE CONSERVATION OF AQUATIC RESOURCES THROUGH MANAGEMENT OF GENETIC RISKS

The relative risks associated with various human activities must be understood to manage genetic resources. The protection of species from human activities that impose only minor risks to genetic resources makes little sense if the species is exposed to other, greater risks concurrently. Human interventions, such as the introduction of the Nile perch into Lake Victoria (see Chapter 2), that impose a high probability of extinction of genetic resources are unacceptable. Other activities, such as regulated harvesting, which might reduce numerical productivity but are less likely to deplete genetic resources, are more acceptable to management.

As Chapter 2 illustrated, commonly encountered human interventions can expose aquatic species to substantial risks of losing genetic resources. However, some human activities expose aquatic species to greater risks than do others because of different genetic processes that may be operating (e.g., founder effects, random genetic drift, selection, migration, and mutation). Thus, human activities that expose populations to random genetic drift or novel selection pressures impose potentially greater genetic risks than activities that involve mutation or migration. Table 3.1 summarizes the genetic processes and relative genetic risks associated with human interventions.

To analyze the potential risks associated with various human interventions and rationalize management programmes based on scientific principles, one must understand the nature of the resources, characteristic components of diversity of the species, genetic processes influencing diversity, and human activities that may place genetic resources at risk of loss. These four groups of considerations are illustrated in Figure 3.1. The relative importance of the factors contributing to genetic risk varies from one species to another. The balance of these for a given species can help to identify the management actions that would minimize the risks of losing genetic resources. Ideally this balance of risk should be quantified, if the relative assessment is to be effective, but in the absence of adequate data on the genetic diversity of the resource at risk, quantifying such risks is highly subjective.

The considerations of Table 3.1 are not uniform for all aquatic species, nor do they arise evenly over time. Therefore management programmes that are intended to minimize the risk of losing genetic material must be tailored to the nature and degree of risk at a given time. Thus the requirements for conserving genetic resources in aquatic animals must be considered on a species-by-species, case-by-case basis. Although there are many common considerations, no universal rules exist for managing the genetic resources of aquatic species.

Table 3.1. Genetic processes and relative genetic risks associated with human activities.

Activities and their effects	Genetic processes	Relative genetic risk[1]
Modification of environments:		
Toxic pollution	Drift, increased inbreeding, selection, extinction	High to extreme
Mutagenic pollution	Increased mutational load	Low
Conditions near limits of physiological tolerance	Drift, increased inbreeding, selection	High to extreme
Harvesting and culturing:		
Overharvesting	Drift, extinction	High to extreme
Selective harvesting	Selection	Moderate to high
Overharvesting small or productive stocks in mixed stock or species fisheries	Selection, drift, extinction	High to less extreme
Overharvesting of natural stocks in culture-based fisheries	Selection, drift, migration	Moderate to extreme
Interbreeding of hatchery and wild stocks	Migration	Low to moderate
Small founder populations for hatchery stocks	Founder effects	High
Too few breeders in cultured stocks	Drift, inbreeding	High
Transfers, introductions and enhancement programmes:		
Introduced exotics alter communities and food webs	Drift, selection, extinction	High to extreme
Interbreeding of stocked and wild populations	Migration	Low to moderate

[1]The risk of losing genetic resources is relative to other interventions. The scale indicates low probability of loss of genetic resources (low); intermediate risks (moderate and high); and risks of extinction (extreme).

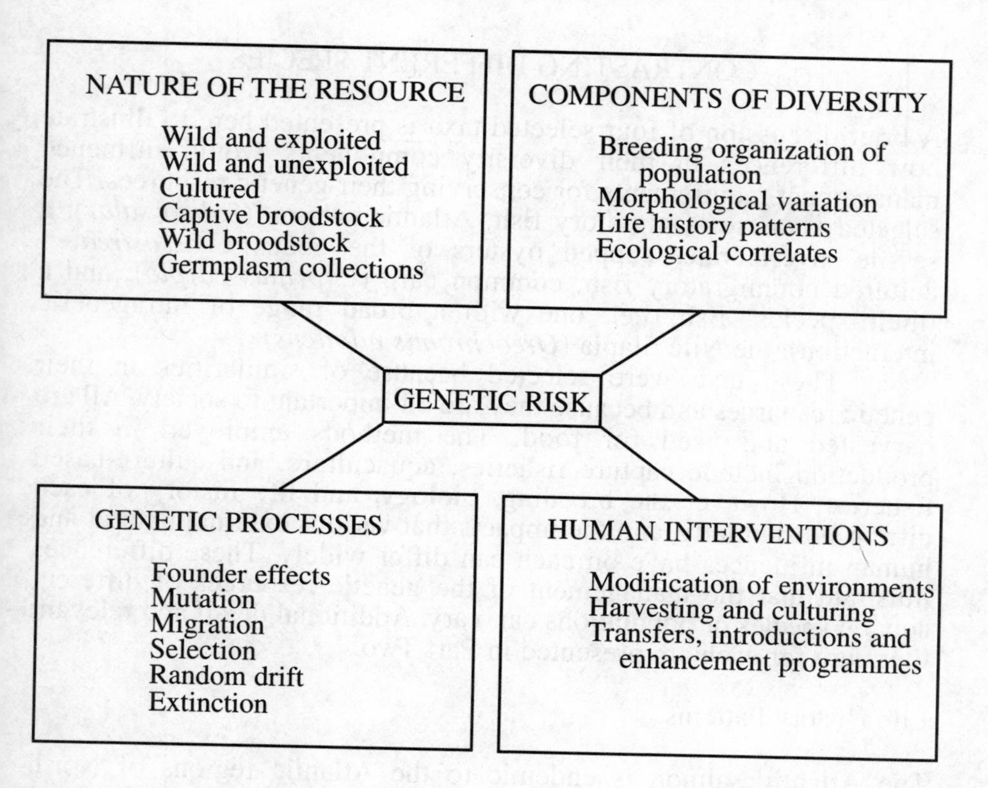

Figure 3.1. Considerations for determining genetic risks to aquatic species.

However, the matrix approach of Fig. 3.1 gives rise to the following questions:

- What is the nature of diversity in this species?
- What are the human interactions with this species?
- What are the biological consequences of these interactions for the species?
- How will these affect the maintenance of its genetic diversity?
- In view of this, what steps should be taken to avoid loss of genetic diversity?
- What field tests are needed to evaluate the effectiveness of these measures?

All these questions should be asked when considering management action for conserving genetic resources of a given species.

The highly disparate level of knowledge about the genetics of aquatic species aggravates the management problem. The relationships between the components of diversity and genetic risks among taxa may be illustrated by contrasting the characteristics of different taxa.

CONTRASTING DIFFERENT SPECIES

A brief discussion of four selected taxa is presented here to illustrate how differences in their diversity components would influence management requirements for conserving their genetic resources. The selected taxa are a migratory fish, Atlantic salmon (*Salmo salar*); a sessile invertebrate, cupped oysters of the genus *Crassostrea*; a cultured nonmigratory fish, common carp (*Cyprinus carpio*); and a 'multi-species' fish (i.e., one with a broad range of intrageneric interaction), the Nile tilapia (*Oreochromis niloticus*).

These taxa were selected because of similarities in their genetic resources and because they are all important to society. All are harvested and used for food. The methods employed in their production include capture fisheries, aquaculture, and culture-based fisheries. However the breeding, biology, and life history of each differs substantially, and the impacts that various external natural and human influences have on each can differ widely. These differences illustrate that the management of the genetic resources of different aquatic species or populations can vary. Additional detail and relevant literature for each are presented in Part Two.

Life History Patterns

The Atlantic salmon is endemic to the Atlantic regions of North America and Europe. Most Atlantic salmon are migratory (anadromous), although resident populations are not uncommon. Atlantic salmon spawn in streams and rivers draining into the North Atlantic Ocean. The 100 to 30,000 fertilized ova are buried in the stream-bed gravel where they incubate unattended for several months. Fully developed fry emerge from the gravel and begin feeding on the aquatic fauna of the stream or river. They show wide developmental variation, comparable to that of several Pacific salmon species (*Oncorhynchus* spp.), but in addition they are iteroparous. After 1 to 8 years of growth in fresh water, the juveniles undergo physiological changes associated with downstream migration and entry to the ocean that enable them to acclimatize to sea water. At this stage they enter the ocean, where they remain for 1 to 4 or more years and undertake lengthy migrations. At the conclusion of the ocean phase, adults re-enter rivers to begin another reproductive cycle. Atlantic salmon show a remarkable propensity for returning to their natal rivers and streams, with only a minor proportion entering systems other than those where they originated.

The cupped oysters are benthic, intertidal, marine species that are widely distributed throughout the temperate and subtropical oceans of both the northern and southern hemispheres. All species are sessile in adult stages. Oysters have exceedingly high fecundities: an adult oyster may release more than 100 million ova in a single spawning season. Oyster populations initiate mass spawning in response to environmental stimuli that correspond to favourable conditions for larval growth and development. Fertilization is external, and the zygotes develop into free swimming planktonic larvae within a few hours of fertilization. The planktonic larval phase lasts from two to several weeks depending on environmental conditions. The larvae are dispersed by tidal currents. At the conclusion of this phase the surviving larvae attach to solid substrates and metamorphose into the sessile life stage. Oysters are filter feeders throughout their lives. In favourable environments they reach sexual maturity in the second year of life and are reproductively active throughout their life span of 10 to 20 or more years.

The common carp is thought to have evolved in Asia Minor, but has been transferred to and has colonized successfully in many areas of the world. Its omnivorous feeding habits and tolerances to a wide range of environmental conditions have enabled it to adapt to both temperate and tropical climates. Among fish species, carp are relatively fecund, with a mature female producing from 100,000 to more than 1 million ova per season. The externally fertilized, adhesive eggs are deposited on aquatic plants or submerged debris at the time of spawning. The embryos hatch within days of spawning, and the free swimming larvae begin feeding shortly after hatching. Neither the incubating eggs nor the emergent fry are attended by parents. Common carp grow rapidly and may reach sexual maturity in the first year in warmer climates, but may not become sexually mature until age 3 or older in cooler climates.

The Nile tilapia is another freshwater species that has been introduced widely around the world. Indigenous to the Nile basin and across equatorial Africa from approximately 5° to 15°N, *O. niloticus* has been introduced now to every continent. Like common carp, Nile tilapia are herbivores that can adapt to a wide range of environments, but unlike common carp, reproduction in *O. niloticus* is much more specialized. At the time of spawning the externally fertilized ova are deposited in hollow nests constructed by the males. A typical spawning results in the deposition of 50 to 2000 eggs during a period of one to several hours. The parents protect the eggs aggressively during the spawning period. Following egg deposition, the female removes the eggs from the nest and carries them in her mouth for 10 to 20 or more days. Males may defend territory for the females during this mouth-brooding period, which enhances the young's chances of survival. Fecundity is low to moderate, with typical values ranging from 500 to 5000 eggs per female per season, deposited during several spawning and brooding events.

Population Structure

The more that a species is subdivided into reproductively isolated units, with minimal genetic interchange, the more it is at risk to loss of genetic diversity from environmental change and from human interference. The four selected examples range from a highly stock-structured species (Atlantic salmon) to an almost panmictic one (cupped oysters), through intermediates of more physiological plasticity (carp and Nile tilapia). This classification is not comprehensive, as the oysters show local physiological differences, and the salmon show developmental plasticity. However, the selection covers a range of adaptive specialization within commonly exploited species, and thus within potentially genetically threatened ones.

Based on observations of allelic variation as revealed by electrophoresis, Atlantic salmon are clearly divided into identifiable breeding populations. There is an apparent hierarchical differentiation from intercontinental through regional to drainage basin and within basin populations. In general the genetic similarity in allele frequencies tends to decrease as geographical distance between populations increases. The pattern is consistent with contemporary views on the evolution of modern salmonids. Small founding populations are thought to have colonized the river basins following the recession of glaciation. Limited (but significant) gene flow, related to anadromous behaviour and a strong tendency to reproduce in natal streams, followed colonization. Natural selection has probably operated to maintain the timing of critical physiological and reproductive events to correspond with local climatic seasons.

In the cupped oysters, geographic variation in allele frequencies is evident only over large distances. For example, there are four prefectural races of the Pacific oyster, *Crassostrea gigas,* over the species' natural range from the north islands to the south of Japan. The races have distinct morphological and physiological characteristics, but are similar in patterns of allelic variation. Similarly, the American oyster, *C. virginica,* which is found along the Atlantic coast of North America from Canada to the Caribbean, exhibits pronounced clines in physiological characteristics over the range. Electrophoretic analysis of allelic variation indicates only four major groupings, and numerous studies have failed to detect significant differences in gene frequencies among or between bays and estuaries over distances of several hundred miles. However, restriction fragment length polymorphism (RFLP) studies of oyster mitochondrial DNA all along the coast from the Gulf of St. Lawrence to Brownsville, Texas, have shown a discontinuity in groupings of clones north and south of a region on the mid-Atlantic coast of Florida, estimated to differ in nucleotide sequence by about 2.6%.

Although common carp from different geographic regions are known to have distinct morphological and physiological characteristics, information about the genetic structure of natural populations is scarce. However, years of experience in breeding domestic strains have shown that those from different geographic regions exhibit substantial genetic differences. Varying degrees of heterosis have been observed when cross-breeding geographic strains. Whether the observed heterosis is related to evolutionary differentiation, or if the additive genetic variance of domestic strains has been exhausted through many generations of intensive selection, is not clear.

Genetic variation in cultured populations of *O. niloticus* has been studied intensively. However, virtually nothing is known about the genetic structure of natural populations within their native range. Investigators have described several morphologically distinct forms, considered to be allopatric subspecies, within the natural distribution of the species. They hybridize readily with other tilapiine species, both in culture and when introduced as an exotic species. Electrophoretic investigations of cultured populations do not reveal substantial genetic variation. Inferences about natural populations cannot be derived from the observations, because nearly all of the cultured populations in the world today were developed from limited numbers of common ancestors. By contrast with Atlantic salmon, where physiological plasticity exists between individuals and between stocks, individual tilapia possess a wide physiological repertoire, and so tolerate extensive environmental changes. However, that repertoire has limits, as shown by the failure of the species to establish self-maintaining populations after release to the wild in Thailand.

While biochemical genetic diversity is very high in oysters (17%), in Atlantic salmon (4%) it is slightly below the vertebrate average (5%), and even below average for anadromous fishes (5.7%) (Ward *et al.*, 1994). The extent of expressed genetic variation (as opposed to phenotypic variation) is not known widely in any of the four examples, but this knowledge is increasing rapidly through breeding experiments. Components of variance between treatments in such experiments have shown genetic influence over developmental rates in all the species. Environmental conditions have been shown to be critical at specific times in determining these rates in salmon, and in determining the sequence of sex changes in oysters. Hence, knowledge of such characteristics is vital for interventive management of these species.

Genetic Risks

The brief sketches contrasting diversity components among the example species may be used to demonstrate how each species might be affected genetically as a result of human activity. The vulnerability of each species depends on the degree to which these impacts affect the genetic processes which restrict diversity, namely drift and selection. Thus the salmon is relatively vulnerable, since many of its stocks are numerically small, its fecundity is modest, and dispersal and gene-flow are minimal. These disadvantages are compensated partly by overlapping generations and a moderate generation time. Oysters are less vulnerable, having high density, high effective numbers of spawners, high fecundity, overlapping generations, a long generation time, and wide dispersal. However, instability of sex ratios and variation in reproductive success owing to temperature and currents (larval survivals are only about 0.0001% to settlement) counterbalance these advantages, and the unexplained excess of homozygotes implies that some selective processes are acting.

As noted in Chapter 2, the main categories of impact are environmental change, fisheries, and introductions.

Environmental Change

All the species are vulnerable to genetic depletion from alterations of their environments. Exposure to toxic pollutants, for example, would impose high to extreme genetic risk to each, probably least to tilapia. However, other forms of environmental modification would impose greater genetic risk to some species than to others. For example, human activities that accelerate the deposition of silt in surface waters would present high to extreme risk to Atlantic salmon and cupped oysters, but a lower degree of risk to common carp and Nile tilapia. Deposition of silt in rivers and streams reduces the area available for salmon spawning, suffocates eggs incubating in the gravel, and affects the benthic food organisms on which juvenile salmon depend. Transport of silt into estuaries places sessile oysters in jeopardy of burial and suffocation. Loss of habitat implies loss of that part of the genetic diversity of the stock which would have used it. By contrast, common carp and tilapia have evolved in silt-laden environments, and their reproductive adaptations enable their survival in waters with heavy silt deposits. Thus, the same form of human intervention would require stringent management in the case of salmon and oysters, but a lesser level of concern for tilapia and carp.

Fisheries

Each of the species would be vulnerable to genetic depletion from overharvesting or selective fishing practices. However, their vulnerability to mixed stock or mixed species fisheries is not uniform. Because of their life history and population structure, cupped oysters would not be exposed to such fisheries. However, Atlantic salmon are extremely vulnerable to the mixed stock fishing problem because of their population structure and migratory behaviour. Natural populations of Nile tilapia are at high risk of genetic depletion in mixed species fisheries because of their ecological interactions as part of multispecies complexes.

Fisheries on all these species select for large size. This has potential for altering life-history strategy, with subsequent generations breeding sooner at a smaller size. Riddell (1986) has argued that this may not reduce genetic diversity, since its greatest effect will be to alleviate density-dependent influences on juveniles. However, recent evidence from Kamchatka (summarized in Thorpe, 1993) has shown that while reducing the mean size and age of a sockeye salmon (*Oncorhynchus nerka*) stock, a prolonged oceanic fishery has also changed the genetic structure of that stock. Large size is associated with relatively slow development and with low levels of heterozygosity. Fish maturing rapidly at very small size tend to show high levels of heterozygosity. The males in this Kamchatka stock are now dominated (>90%) by small residents that are highly heterozygous. As survival probability is highest at modest levels of heterozygosity, as shown experimentally in pink salmon (*O. gorbuscha*) (Altukhov & Salmenkova, 1991), such distortions of the genetic structure of the stock imply a decrease in its stability.

Introductions

Direct ecological effects through competition by introduced non-native stocks may reduce production, as in overplanting on oyster beds, where this can lead to reduction of phytoplankton production and consequent oyster growth, lowering yield. At worst, such overplanting may reduce the remaining native stock through toxic effects of oyster wastes. Transfers have sometimes resulted in reduced genetic diversity through hatchery bottleneck effects, as in US Pacific coast stocks of oysters, which have been swamped by massive plantings of hatchery material produced from too few parents. Tilapias have been transferred and planted all over the warm-water world, creating problems of introgression, hybridization, and competitive replacement of native species. In salmon, the perceived danger is more from massive escapes or deliberate releases of farmed fish of genotypes not locally adapted (Hansen *et al.*, 1991). Such a threat has yet to be established objectively. Two recent modelling studies using realistic assumptions (Hutchings, 1991; Mork, 1991) have estimated the numerical and the genetic effects of such interactions, but the problem urgently requires experimentation (Saunders, 1991; Thorpe, 1991). However, the loss of several stocks of salmon in Norway, through catastrophic exposure to an unfamiliar parasite, *Gyrodactylus salaris*, represents an indirect genetic effect of introducing parasite-tolerant carrier stocks from outside Norway into parasite-intolerant Norwegian populations (Bakke, 1991; Johnsen & Jensen, 1991).

PREPARING FOR CRISES

If management of aquatic resources fails to conserve genetic resources, more extreme actions will be required to preclude further genetic deterioration, or even extinction. These actions should not be considered as alternatives to a management approach that incorporates genetic conservation. Need for these indicates that conservation management has been inadequate to protect the resources. Their intention is to impede further loss.

In Situ Management

In a recent study, the Office of Technology Assessment of the US Congress (Office of Technology Assessment, 1987) concluded that the management options available to maintain biological diversity could be classified conveniently according to the degree of human intervention (Table 3.2). The first level of classification is based on whether management is applied to natural (on-site) or artificial (off-site) environments. On-site management systems include ecosystem maintenance and species management; they are referred to as *in situ* systems in this book. Off-site management systems involve increasing degrees of environmental control and include living collections and germplasm storage; they are referred to as *ex situ* systems in this book.

Ecosystem Maintenance

Ecosystem maintenance involves establishing a defined geographical area that can be managed to insulate it as much as possible from external influences. Although maintenance may involve supplementing some species to maintain minimum viable population sizes, the diversity of these ecosystems evolves primarily in response to natural processes. The ecosystem's genetic resources are not managed as such, but evolve naturally within it.

The population dynamics of any given species in the ecosystem are related to those of other species. Consequently fluctuating populations at any level of ecosystem organization are likely to result in attendant responses at other levels. The ecosystem must also be large enough to provide adequate habitat if a portion of it is lost through natural perturbations, such as volcanic eruptions, floods, fires, and earthquakes. Migratory species or species that depend on gene flow between distinct breeding populations to maintain their evolutionary capacity may be at risk of depletion in restricted or limited ecosystems.

On a smaller spatial scale, specific critical habitats, such as the spawning grounds for salmon, may be restored physically after damage through gravel excavation, or substituted by specially constructed spawning channels to replace lost habitat. Where dams have obstructed

Table 3.2. Examples of management systems to conserve aquatic genetic resources
(from Office of Technology Assessment, 1987).

In situ management systems		*Ex situ* management systems	
Ecosystem maintenance	Species management	Living collections	Germplasm storage
National parks	Wildlife refuges	Aquaria	Semen, ova, embryo banks
Wilderness areas	Hatcheries	Captive broodstocks	
Marine sanctuaries	Harvest regulations	Stocking, and introduction policies	

upstream passage of adult salmon, ladders, ramps and lifts of various kinds have been installed to assist fish to overcome the obstacles. In view of the subdivided nature of salmon stocks into components associated with particular tributaries of individual rivers, such structures are vital to ensure the maintenance of genetic diversity in these systems.

The surrounding region influences environments in many ecosystems. Aquatic ecosystems are especially vulnerable to modification by external human activities. Although some aquatic systems at higher elevations may include their source waters, and consequently may remain relatively pristine and free from outside influences, this may not be the case with lakes and rivers at lower elevations, estuaries, and marine ecosystems. Airborne pollutants (e.g., acid rain) from outside can also result in major alterations of environments within systems.

Species Management

Species management is managing the population size and structure of selected species as opposed to managing their habitats. Familiar examples include wildlife refuges, game parks, and reserves. With the use of artificial and natural propagation between semidomesticated species and their wild relatives, species management helps maintain genetic variation.

Species management has been used to maintain wild populations for sustainable exploitation, and in attempts to maintain minimum viable populations of threatened or endangered species. Fish and shellfish hatcheries are examples, although hatchery management practices are based more on preservation of numbers for harvesting than on genetic conservation.

Ex Situ Management

The technology exists to propagate living collections of many important aquaculture species, and long-term storage of frozen gametes, and even zygotes, may be practical in the near future (Stoss, 1983). Although fish gene banks are feasible, ensuring adequate representation of population gene pools in the collections will be a major undertaking. With many species of fish and shellfish, spawning occurs over weeks or months. Representative sampling of the genetic diversity of populations requires sampling throughout their geographic range and spawning period, and subsequently maintaining a large brood-stock or germplasm collection.

Living Collections

The best known example of living collections are zoological and botanical gardens, fish hatcheries, and aquaria. Many living collections have been developed for research, education, or display purposes. Living collections can also be a useful way to preserve genetic material, including maintaining breeding stocks of populations threatened in the wild.

Living collections require intensive breeding management. With judicious breeding management, collections can limit the amount of inbreeding and random genetic drift relative to that which would occur in small, unmanaged populations. For this reason living collections are an attractive alternative for preserving extremely scarce genetic resources.

Living collections can maintain genetic materials that cannot be stored in gene banks. Currently, technologies for long-term storage of ova or embryos do not exist for many species of aquatic organisms. Consequently the maintenance of living collections is the only controllable alternative for the genetic conservation of some aquatic species, especially viviparous and ovoviviparous species.

Germplasm Storage

Germplasm storage involves the long-term storage of gametes, embryos, or selected tissues, and requires special technologies. The technology for long-term storage of fish sperm has been demonstrated (Stoss, 1983). However, it has not yet been applied broadly as a conservation methodology. Until *ex situ* technologies become routine, storage of tissues from extremely threatened species should be undertaken to prevent complete loss of the genetic information.

CONCLUSIONS

Ideally, resource management policy embodies two goals: conserving the resource, and using the resource. Often, but not necessarily, the two goals are mutually exclusive if conservation is intended to prevent outside forces from changing the genetic make-up of stocks and the inherent variability among stocks within the resource. Clearly exploiting a resource without imparting some genetic change is nearly impossible. The difficult position of resource managers results from the need to achieve sustainable genetic continuity while balancing social and economic pressures.

What may not be obvious is the requirement for having reliable information when making management decisions. Managers are faced with a multitude of questions about genetics, population structure, and the impact of various interventions, of which few can be answered. A considerable body of research will be necessary to gather data needed to obtain answers. Until then, management practices must be based on the understanding gained from a few well-studied cases.

RECOMMENDATIONS

The conservation of aquatic animal resources must address the unique characteristics of individual species and populations.

As the examples of the effects of human activities have illustrated, management actions to conserve genetic resources in aquatic animals must be considered on a species-by-species, case-by-case basis. Managers must consider not only the species' components of diversity and the effects of human interventions, but also the nature of the resources and their use. There is a continuing conflict between conservation and exploitation in aquatic resource management. The conflict arises because aquatic resources are mainly common property resources, and, with forests, they are among the few renewable resources subject to harvest and culture activities.

In the absence of information, fisheries and aquatic resource practices should be based on sound scientific principles that give priority to conservation over exploitation.

A salient feature of the challenges resource managers must address is that most deal with or affect genetics. It is an inescapable conclusion that biological resource conservation is actually genetic conservation, and thus, the application of genetic principles is essential. Unfortunately genetic principles have been ignored largely throughout the history of fisheries and aquatic resource management.

For many endangered species and populations, the development and application of ex situ *methods for preserving tissues, ova, sperm, or whole organisms are needed to supplement, or be an alternative to,* in situ *conservation.*

The inability of most aquatic organisms to store ova and embryos is a serious constraint to applying *ex situ* methods to conservation of aquatic species. For some species, such as endangered cichlids, captive maintenance and breeding may be necessary to preserve the small remnant that remains. *Ex situ* methods, while not preserving the natural adaptive genetic and evolutionary processes, may provide a needed insurance against complete loss of many highly endangered species. Once technologies are available, effective *ex situ* management will require secure, long-term funding.

Although psychologically satisfying and popular, *in situ* conservation may not be achievable easily for all species or populations. While protection of a lake or stream may be possible, it may be more difficult in a marine environment to prevent the inflow of toxic chemicals or pollutants. For migratory or wide-ranging species, protection afforded in one area may be negated by its lack elsewhere. Even where it is possible, establishing ecosystems of sufficient size and diversity to maintain genetic resources may introduce formidable political and economic problems, primarily because of a lack of knowledge about effective size and design. Whether a single ecosystem large enough to maintain species diversity and viable populations of all the species is more effective than several smaller systems is not always clear. Clearly, for many aquatic species *in situ* conservation will be effective only if coupled with appropriate international cooperation to conserve and manage the resources responsibly throughout their range.

4
STATUS OF AQUATIC GENETIC RESOURCE MANAGEMENT

Management programmes that prevent the deterioration of environments and regulate harvesting, culturing, and transfers and introductions can reduce the genetic risks to many aquatic species. Substantial progress has been achieved in this regard in recent years. International cooperation, manifested in a proliferation of treaties and international fisheries commissions, is a positive trend in addressing conservation needs. Similarly, national and regional policies and management plans have increased in both frequency and sophistication, but may prove to be too limited and too late to ensure the survival of some aquatic species. Conservation of aquatic genetic resources must become a national and international priority, to avert the loss of many important aquatic species.

This chapter reviews conservation activities from a global perspective, with emphasis on the activities of international organizations and the status of conservation programmes for selected economically important species.

INSTITUTIONAL CONSTRAINTS

Many nations have some formal capacity for administering aquatic natural resources. These resources and issues of their use are highly complex, and many countries have developed correspondingly complex administrative and institutional structures for aquatic resources management. Each agency (e.g., department, ministry, office) has its own mandate and agenda. The resultant partitioning often results in overlapping, and even conflicting, jurisdictions that limit the ability of any office to satisfy its mandate. For example, a government office dealing with fish and shellfish resources may develop management plans for harvesting, culturing, transferring, or enhancing these species, but be incapable of protecting them adequately because it has no jurisdiction over the environments in which they occur.

Separate agencies are likely to administer management programmes directed toward different societal activities, all of which may affect a common aquatic resource community. The agency responsible for protecting fish and shellfish may have no authority to regulate such activities. In some cases the human activities may occur in aquatic environments, but unrelated land-based activities also can have substantial impacts on aquatic or oceanic environments. These activities include thermal and hydroelectric power generation; petroleum or mineral exploration and extraction; agriculture, forestry, and rural development; transportation along or construction of highways, railways, and marine shipping lanes; and industrial or urban development, including water supply and waste disposal.

These land-based activities can produce alterations in aquatic environments, such as temperature effects, pollution, barriers to migration, and deterioration of habitats. In some cases, ecosystems are so altered as to have lasting impacts on the genetic resources of aquatic animal populations, as, for example, the total loss of fish populations (and in some cases fish communities) in Canada, Britain, and Scandinavia as a result of reduction in pH due to acid rain (Jensen & Snekvik, 1972; Schom & Davidson, 1982). For most species, however, the significance of the loss or alteration of the environment is unknown. Nevertheless, the loss of potentially useful and ecologically important species must be addressed aggressively if further irreparable damage is to be prevented. The most limiting constraints to taking appropriate action are a severe lack of knowledge about aquatic species and their interactions with their environments (addressed earlier), and the inadequacy of national and international institutional and legal systems to implement appropriate action programmes. To remedy the problem of lack of knowledge, the International Centre for Living Aquatic Resources Management (ICLARM), together with the UN Food and Agriculture Organization (FAO) have started a database on living aquatic resources, known as FishBase (see Box 2).

INTERNATIONAL FISHERIES ORGANIZATIONS

Many international bodies are involved actively in conserving aquatic species. Some of them have been established by conventions or treaties among nations, and are empowered to enact fisheries management policies. Most countries are now signatories to one or more international conventions or resolutions. However, many organizations are advisory and have little authority to establish policy or enforce it.

Organizations Established by International Conventions

Through this century international conventions and treaties have established several international fisheries commissions for conserving certain species (Table 4.1). The jurisdictions of these commissions extend to most of the world's surface waters and are directed toward many species of economic importance. International commissions were formed to promote the conservation of a single species or group of species exploited in international waters by two or more nations. Commissions have evolved organizational structures that serve their unique needs.

FishBase

Effective resource management depends on the availability of accurate information for researchers, planners and managers. In the case of fisheries and aquaculture, the information available is dispersed widely in textbooks and thousands of journal articles, with little standardization of terms and measurements used. Bibliographic databases, such as the Aquatic Sciences & Fisheries Abstracts, provide access to key terms and abstracts, but researchers must find the original literature for details of any use in comparative studies.

To remedy this, the International Centre for Living Aquatic Resources Management (ICLARM) joined with the UN Food & Agriculture Organization (FAO) to start a database on living aquatic resources, known as FishBase. The initial reason for setting up this system was the need for data on structured population dynamics. The new resource aims to summarize global information on fish crustaceans and molluscs in a standardized form, and uses high-quality colour pictures to allow quick, easy, accurate identification.

By 1994, FishBase contained information on about 12,000 fish and crustaceans. Included are scientific names and synonyms, FAO names, common names by country, economic importance by species, and species distribution by FAO areas and countries.

About 150 important diseases for 38 fish species have been described also, including their prevalence, symptoms, effects, treatments and prevention methods. Built around a regional database, the system provides a way to make comparisons between species groups and geographical areas. As a research tool, it is the first system to provide sufficient information for such invaluable comparative studies.

Details on abundance, habitats, behaviour, reproduction, food items and consumption, predators, competitors, and status regarding threats are expected to help identify endangered fish species and habitats, and hence to aid attempts to conserve biological diversity. This is particularly valuable because technical difficulties and cost preclude the maintenance of large-scale collections of fish germplasm. With knowledge of which species are most threatened, special efforts can be made to preserve germplasm of particular species.

Another important aim of FishBase is to provide complete

morphological information on fish. Description of the size and form of all fish larvae in the North Sea and Mediterranean have already been entered into the system. The inclusion of drawings and descriptions from publications dating back to the last century allows researchers to track species evolution and geographical spread.

ICLARM staff in Manila, the Philippines, are responsible for most of the data entry on FishBase. Information is also received from regional outposts in the Philippines and Malawi, and there are cooperation agreements with experts and museums in Belgium, Germany, and the Netherlands. At the moment, the system is funded principally by the European Union, with additional support from FAO for information on shrimps and molluscs.

Given its aim of providing information for developing countries, one of FishBase's most important features is that it is designed to run on low cost IBM compatible microcomputers. The complete data base will be stored on CD-ROM disks, with updates expected annually. Regional or national data will be available on standard diskettes and, where appropriate, in hard copy. The first version of FishBase was released for comment to its donors and collaborators in 1994.

[From Froese (1990), *Naga* (1993) and A.Eknath (pers.comm. 1994).]

Most commissions have standing committees or panels to address resource allocation, enforcement, research, and administration.

Although these commissions generally were not founded to conserve genetic resources, most now recognize their importance. Several have promoted research to provide information for genetic resource management. To the extent that the commissions are empowered to negotiate and enforce resource management policy, they may represent the most viable approach to the conservation of genetic resources of species harvested in international waters. International organizations also could promote and support collaboration among nations on research issues of global significance.

Advisory Organizations

Advisory organizations play an important role in conserving fishery resources by facilitating discussion and fostering cooperation among nations. The following are examples of a few organizations to illustrate the diversity of these groups and their activities.

IUCN: International Union for the Conservation of Nature and Natural Resources

The International Union for the Conservation of Nature and Natural Resources was founded in 1948 and claims to be the largest professional body in the world working to care for the soils, lands, waters, and air of the planet and the life they support (Anon., 1990). In 1980, IUCN published the *World Conservation Strategy* in collaboration with the United Nations Environment Programme (UNEP) and the World Wide Fund for Nature (WWF). This important document was one of the first to promote sustainable development, a concept widely hailed since it was endorsed by the World Commission on Environment and Development in 1987. The *Strategy* set out three basic principles for conservation: that essential ecological processes and life-support systems must be maintained; that genetic diversity must be preserved; and that any use of species and ecosystems must be sustainable. To promote movement towards these goals, IUCN has worked with more than 30 governments to develop national conservation strategies.

In 1988, work began on a new version of the *Strategy*, again in collaboration with UNEP and WWF. Through a broad-based consultation process and workshops on specific topics around the world, the three partners developed *Caring for the World: A Strategy for Sustainability* (see Box 3). Released in 1991, the new strategy contributed to the deliberations leading up to the 1992 UN Conference on Environment and Development held in Rio de Janeiro. In the two chapters on aquatic resources, the strategy focuses on conservation of genetic diversity through ecosystem and habitat management.

FAO: Food and Agriculture Organization

The Food and Agriculture Organization of the United Nations has been active in promoting the conservation of fishery resources since its inception. Several international fisheries advisory bodies have been established by FAO resolutions (Table 4.2).

In cooperation with the United Nations Environment Programme (UNEP), FAO convened and published the proceedings of an Expert Consultation on the Genetic Resources of Fish in 1980 (Food and Agriculture Organization, 1981). The FAO, the International Council for Exploration of the Sea, and the American Fisheries Society jointly

Table 4.1 International fishery commissions established by conventions and treaties.

Year	Commission	Member Nations
1902	International Council for the Exploration of the Sea (ICES)	Belgium, Canada, Denmark, Finland, France, Germany, Iceland, Ireland, Netherlands, Norway, Poland, Portugal, Russia, Spain, Sweden, UK, USA.
1923	International Pacific Halibut Commission (IPHC)	Canada, USA.
1937	International Pacific Salmon Commission (IPSC) (replaced by treaty in 1985)	Canada, USA.
1946	International Whaling Commission	Argentina, Australia, Brazil, Canada, Denmark, France, Iceland, Japan, Mexico, Netherlands, New Zealand, Norway, Panama, Russia, South Africa, UK, USA.
1949	International Commission for Northwest Atlantic Fisheries (ICNAF)	Canada, Denmark, France, Germany, Iceland, Italy, Japan, Norway, Poland, Portugal, Romania,Russia, Spain, UK, USA.
1949	Inter-American Tropical Tuna Commission (IATTC)	Costa Rica, USA, Panama (1953), Ecuador (1961) Mexico (1964), Canada (1968), Japan (1970), France (1973), Nicaragua (1973). subsequently withdrew: Ecuador (1968), Mexico (1978), Costa Rica (1979), Canada(1984)
1952	South Pacific Permanent Commission (SPPC)	Chile, Colombia, Ecuador, Peru.
1952	North Pacific Fur Seal Commission (NPFSC)	Canada, Japan, Russia, USA.
1952	International North Pacific Fisheries Commission (INPFC)	Canada, Japan, USA.

1959	Mixed Commission for Black Sea Fisheries	Bulgaria, Romania, Russia.
1960	Indo-Pacific Fishery Commission (IPFC)	Australia, Bangladesh, France, India, Indonesia, Japan, Korea, Malaysia, New Zealand, Philippines, Sri Lanka, Thailand, UK, USA.
1962	Joint Fishery Commission (JFC)	Bulgaria, Cuba, Germany, Poland, Romania, Russia.
1966	International Commission for Conservation of Atlantic Tunas (ICCAT)	Angola, Brazil, Canada, Côte d'Ivoire, Cuba, France, Gabon, Ghana, Japan, Korea, Morocco, Portugal, Sénégal, Russia, South Africa, Spain, USA.
1969	International Commission for Southeast Atlantic Fisheries (ICSEAF)	Angola, Bulgaria, Cuba, France, Germany, Iraq, Israel, Italy, Japan, Korea, Poland, Portugal, Romania, Russia, South Africa, Spain.
1973	International Baltic Sea Fishery Commission (IBSFC)	EU, Finland, Poland, Russia, Sweden.
1978	Northwest Atlantic Fisheries Organisation (NAFO)	Bulgaria, Canada, Cuba, Denmark, EU, Germany, Iceland,Japan, Norway, Poland, Romania, Russia.
1979	South Pacific Forum Fisheries Agency (SPFFA)	Australia, Cook Islands, Micronesia, Fiji, Kiribati, Nauru, New Zealand, Niue, Palau, Papua New Guinea, Samoa, Solomon Islands,Tonga, Tuvalu, Vanuatu.
1980	International Commission for the Conservation of Antarctic Marine Living Resources (CCAMLR)	Argentina, Australia, Belgium, Brazil, Chile, EU, France, Germany, India, Japan, Korea, New Zealand, Norway, Poland, Russia, South Africa, UK, USA.
1980	Northeast Atlantic Fisheries Commission (NEAFC)	Bulgaria, Denmark, EU, Germany, Iceland, Norway, Poland, Russia, Sweden.

1982	North Atlantic Salmon Conservation Organisation (NASCO)	Canada, Denmark, EU, Iceland, Norway, Russia, Sweden, USA.
1984	Latin American Organisation for the Development of Fisheries (OLDEPESCA)	El Salvador, Guatemala, Mexico, Nicaragua, Panama, Peru, Venezuela.

developed a code of practice and protocols relating to introductions and transfers that were distributed to other commissions and organizations (International Council for Exploration of the Sea, 1984). The code of practice sets out guidelines and a policy statement to supplement local legislation. The protocols are a checklist for implementing the components of the code of practice. The code has no legal status, but its adoption by member countries implies a strong moral obligation to implement the practices, by legislation if necessary. The member countries of the European Inland Fisheries Advisory Commission (EIFAC) have adopted the code, as did the Committee for Inland Fisheries of Africa in 1990. The Comision de Pesca Continental para America Latina (COPESCAL, Commission for Inland Fisheries of Latin America), and the Indo-Pacific Fisheries Commission reviewed the code of practice and adopted it in 1992. The other commissions continue to scrutinize it to determine its applicability to their regions and needs. The FAO has recommended recently (Bartley,1994) that the code should Box 3.

IUCN: *Caring for the World: A Strategy for Sustainability*

In 1980 the International Council for the Conservation of Nature and Natural Resources (IUCN), in collaboration with the United Nations Environment Programme (UNEP) and the World Wide Fund for Nature (WWF), published their *World Conservation Strategy*, which set out their three basic principles for conservation:

- essential ecological processes and life-support systems must be maintained;
- genetic diversity must be preserved;
- any use of species and ecosystems must be sustainable.

In 1988 these three organizations began work on a new version of the *Strategy,* developing *Caring for the World: A Strategy for Sustainability*, which was released in 1991. Two of the seven chapters deal specifically with aquatic resources. In a section on fresh water, the strategy notes that the loss of habitat is endangering several hundred fish and invertebrate species. In addition, the genetic resources used for aquaculture are probably leading to genetic change; as species and genetically distinct stocks are transferred from one region to another, native gene pools are being disrupted. Further, biodiversity is being reduced by changes in water flow that follow the construction of dams.

To address these problems, the strategy calls for an ecosystem approach to water management, based on drainage basins as the unit of management. Inventories are needed on the full range of products and services that are taken from each part of the system. The results could support the passage of water licensing laws that recognize the non-consumptive values of water, such as for recreation and as the site of important genetic resources. Water demand needs to be managed for efficient and equitable allocation among competing water uses. The strategy also calls for research on the genetic diversity of the freshwater species used in aquaculture, and on how far original stocks can or should be maintained.

The chapter on oceans notes that coastal zones have the highest biological productivity on earth, and that coastal waters and the continental shelf yield more than 90% of the world's fish catch. This vital resource is under pressure from both pollution and overfishing. More than 75% of marine pollution comes from land-based sources, as urban and industrial development crowds coastlines and as chemicals used by farmers run off their fields into rivers and eventually to the ocean. The presence of pathogens from sewage has led to some stocks of shellfish being declared unfit for human consumption. Tackling marine pollution problems thus entails major changes in industrial and agricultural practices.

On overfishing, the strategy notes that most stocks of fish that live at or near the bottom of the sea are fully fished. Most fisheries are exploited beyond the levels that are economically optimal, with unknown consequences for the genetic diversity and adaptability of the stocks. *Caring for the World* calls for the recovery of depleted fisheries by the year 2010, and for no overexploitation of fisheries.

One reason cited for the dismal situation of these aquatic resources is the fragmented and sectoral way that problems of coastal and marine ecosystems are addressed. As with freshwater resources, the strategy calls for an ecosystem approach, and for determining the system's carrying capacity.

Several regional efforts at ecosystem management are cited as steps in the right direction - the regional seas programme of UNEP and the work of the Commission on the Conservation of Antarctic Marine Living Resources. Regional action plans need to be developed for the Arctic Ocean, Black Sea, South Asian Seas, North Pacific, Southwest Atlantic, South China Sea, and the Sea of Japan.

National coastal zone management policies should determine how the needs of many sectors could be balanced, and how conservation and development can be combined. Many areas are threatened by sea level rise, storm surges, coastal flooding, or erosion. In almost all cases, the strategy notes, the best coastal protection involves establishing buffer zones where the uses are compatible with natural nearshore processes.

Noting that the worldwide network of coastal and marine protected areas is far less developed than terrestrial ones, the strategy calls for a global system plan to be prepared that would safeguard representative ecosystems and prepare management plans. The full system should be in place by the year 2010. For the vast open ocean beyond the jurisdiction of Exclusive Economic Zones, an effective legal management regime is needed.

[Anon., 1990.]

be made more visible and more practical. To this end they have proposed the compilation of a users manual to be entitled 'Guidelines for the responsible introduction and transfer of aquatic organisms'. The FAO has also published a register of transfers of inland fish species that is being maintained and updated.

Based on recommendations of the Technical Cooperation Network for Aquaculture, organized by the FAO Regional Office for Latin America and the Caribbean, the government of Costa Rica offered facilities to establish a regional gene bank for tilapia species, but this was not set up due to lack of funds. There is now interest in establishing gene banks for conservation in Colombia and Venezuela, but so far only small training courses and demonstrations on spawning, cryopreservation and thawing have been completed under the supervision of the International Fisheries Gene Bank, Vancouver, Canada.

EIFAC: European Inland Fisheries Advisory Commission

The commission was organized in 1957 to promote cooperation among European nations in managing the fisheries resources of the continent's inland waters, especially the resources of river basins shared by two or more countries. In addition to its action on the code of practice, EIFAC

organized a symposium on the new developments in aquacultural genetics in 1986. The symposium made a series of recommendations concerning the genetic bases
of species improvements, selective breeding, hybridization, genetic manipulations, and large-scale breeding programmes.

ICLARM: International Center for Living Aquatic Resources Management

The International Center for Living Aquatic Resources Management (see Box 4) was established in the Philippines in 1976 as an autonomous, nonprofit organization to complement and support the activities of national and regional institutions working in fisheries and aquaculture research and development in developing tropical countries. In 1987, ICLARM organized a workshop on tilapia genetic resources for aquaculture, and published French and English translations of the workshop's proceedings (Pullin, 1988).

With funding from Germany, the Asian Development Bank, and the United Nations Development Programme, ICLARM is implementing a worldwide programme to collect germplasm for establishing base populations of tilapia species. These stock collections, and the improved breeds to be developed from them, are envisaged to be resources of global importance. Collection of germplasm is under way in Egypt, Ghana, and Sénégal.

In 1992 ICLARM developed a strategic plan for international fisheries research, at the invitation of the Consultative Group on International Agricultural Research (CGIAR).

Table 4.2. International fishery commissions established by resolutions of the Food & Agriculture Organisation (FAO) of the United Nations.

Year	Commission	Member Nations
1948	Indo-Pacific Fishery Commission (IPFC)	Australia, Bangladesh, Cambodia, France, India, Indonesia, Japan, Korea, Malaysia, Myanmar, Nepal, New Zealand, Pakistan, Philippines, Sri Lanka, Thailand, UK, USA, Vietnam.
1949	General Fisheries Council for the Mediterranean (GFCM)	Algeria, Bulgaria, Cyprus, Egypt, France, Greece, Israel, Italy, Lebanon, Libya, Malta, Monaco, Morocco, Romania, Spain, Syria, Tunisia, Turkey, Yugoslavia.
1957	European Inland Fisheries Advisory Commission (EIFAC)	Austria, Belgium, Bulgaria, Cyprus, Czechoslovakia, Denmark, France, Finland, Germany, Greece, Hungary,

		Ireland, Israel, Italy, Netherlands, Norway, Poland, Portugal, Romania, Spain, Sweden, Switzerland, Turkey, UK, Yugoslavia.
1961	Regional Fisheries Advisory Commission for the Southwest Atlantic (CARPAS)	Argentina, Brazil, Uruguay.
1967	Fishery Commission for the Eastern Central Atlantic (CECAF)	Benin, Cameroon, Cape Verde, Congo, Côte d'Ivoire, Cuba, Equatorial Guinea, France, Gabon, The Gambia, Ghana, Greece, Guinea, Guinea-Bissau, Honduras, Italy, Japan, Korea, Liberia, Mauritania, Morocco, Nigeria, Norway, Poland, Romania, São Tome & Principe, Sénégal, Sierra Leone, Spain, Togo, USA, Zaire.
1967	Indian Ocean Fishery Commission (IOFC)	Australia, Bahrain, Bangladesh, Comoros, Cuba, Ethiopia, France, Greece, India, Indonesia, Iran, Iraq, Israel, Japan, Jordan, Kenya, Korea, Kuwait, Madagascar, Malaya, Maldives, Mauritius, Mozambique, Netherlands, Norway, Oman, Pakistan, Portugal, Qatar, Saudi Arabia, Seychelles, Somalia, Spain, Sri Lanka, Sweden, Tanzania, Thailand, United Arab Emirates, UK, USA, Vietnam.
1971	Committee for Inland Fisheries of Africa (CIFA)	Benin, Botswana, Burkina Faso, Burundi, Cameroon, Central African Republic, Chad, Congo, Côte d'Ivoire, Egypt, Ethiopia, Gabon, The Gambia, Ghana, Guinea, Kenya, Lesotho, Madagascar, Malawi, Mali, Mauritius, Niger, Nigeria, Rwanda, Sénégal, Sierra Leone, Somalia, Sudan, Swaziland, Tanzania, Togo, Uganda, Zaire, Zambia, Zimbabwe.
1973	Western Central Atlantic Fishery Commission (WECAFC)	Antigua & Barbuda, Bahamas, Barbados, Brazil, Colombia, Cuba, Dominica, France, Grenada, Guatemala, Guinea, Guyana, Haiti, Honduras, Jamaica, Japan, Korea, Mexico, Netherlands, Nicaragua, Panama, St Christopher & Nevis, St.Lucia, St.Vincent & the Grenadines, Spain, Suriname, Trinidad & Tobago, UK, USA, Venezuela.

1976	Commission for Inland Fisheries of Latin America (COPESCAL)	Argentina, Bolivia, Chile, Colombia, Costa Rica, Cuba, Dominican Republic, Ecuador, El Salvador, Guatemala, Honduras, Jamaica, Mexico, Nicaragua, Panama, Paraguay, Peru, Suriname, Uruguay, Venezuela.

The plan called for a new institute to study improvements in fish production and management. ICLARM became a member of CGIAR in 1992 and is playing an even larger role in aquaculture. In particular, it will document, through FishBase (see Box 2), a catalogue of fish genetic resources from freshwater ponds, reservoirs, lakes, estuaries, lagoons, coastal ponds, and coral reefs. Through collaboration with other research organizations (e.g., Stirling and Swansea Universities, UK [cryopreservation, and population dynamics]; Dalhousie University, Canada [molecular genetics]; Szarvas Institute, Hungary [carp breeding]; and Akvaforsk, Norway [breeding programmes]), it is improving fish germplasm for aquaculture (see Chapter 8 below). It will also collaborate with other centres with specialist knowledge of particular species, in India, China, and Africa, and through the Aquaculture Genetics Network in Asia, sponsored by IRDC. It will continue its project on Genetic Improvement of Farmed Tilapias (GIFT), which is showing that use of genetic resources and research towards better breeds can stimulate the initiation of national breeding programs. ICLARM expects that by the end of 5 years their improved *O. niloticus* breeds will be available in most Asian countries that farm the species. Improvements will include a probable 40% faster growth in a variety of low-input systems, and progress in a delayed maturation of production stocks. They also plan a tilapia germplasm reference collection, to be duplicated or maintained by the institute and host country, as a major source of germplasm for further research. Approaches will also be made towards breeding programmes for Chinese and Indian carps (Anon., 1991c).

AWB: Asian Wetland Bureau

The Asian Wetland Bureau is an international nonprofit organization which aims to promote protection and sustainable use of wetland resources in Asia. It works through governmental and nongovernmental bodies throughout S.E. Asia from India to South Korea. Its four main objectives are to:

- minimize loss of wetlands and their biodiversity;
- establish a network of managed wetland reserves;
- ensure sustainable management of these, in harmony with local livelihoods; and
- ensure that national governments in Asia collaborate to manage wetland sites of international importance.

It is surveying the conservation status of freshwater fishes in Malaysia and promoting improvement of the legislation on their use. It is also pursuing an active role in conservation education in the region.

Box 4.

ICLARM: Conserving Aquatic Genetic Diversity

The issue of conserving biological diversity has captured a good deal of media coverage and public interest over the last few years. Attention has focused generally on tropical rain forests and on what are known as the megavertebrates - for example, elephants and the black rhinoceros in Africa. Arresting photographs of the charred remains of a forest in Brazil or of an elephant slaughtered by ivory poachers telegraph the message that species and ecosystems are under threat.

Yet aquatic ecosystems and the species they contain are equally threatened. It is just more difficult to capture their loss in a dramatic photograph or two. In North America, for instance, out of approximately 700 freshwater finfish species, 157 are thought to be threatened with extinction (see Table 1.1).

Aquatic ecosystems are poorly understood, relatively unmanageable, shared by multiple users, and highly vulnerable to human interventions and climate change. The first priority for conservation is better documentation of the status of the genetic diversity of aquatic organisms.

The second key step is to evaluate the contribution of these resources, so that a more accurate picture of the impact of any losses can be formed. The idea that preserving genetic diversity is a moral imperative holds less sway in today's world than the argument that not doing so will lower our quality of life by reducing our options for food, health care, income, and recreation sites. This approach is being used successfully regarding other parts of the earth's biodiversity. Recent studies, for example, have looked at the medicinal and agricultural value of plant genetic resources.

Estimating the value of aquatic genetic diversity is not as easy, however. For commercial and sport fisheries, the costs of overfishing or the expansion of aquaculture is less clear. Cost and technological difficulties preclude the establishment of large-scale collections of fish germplasm. Therefore, with every irreversible genetic change a potential resource for future breeding programmes is lost.

Given the problems of having fish gene-banks that cover all the species and stocks in the world, the protection of entire habitats and ecosystems is even more important for aquatic resources than it is for plants and livestock. Thus documentation of the value of leaving habitats undisturbed because of the genetic resources they contain is a vital research area that needs to be pursued.

Work on conserving the genetic resources of the tilapia that are native to Africa provides an example of one useful approach. Although some researchers claimed that the development of this species for aquaculture has been so widespread that no undisturbed habitats and populations still existed, a 1987 meeting organized by ICLARM did identify some undisturbed sites.

Several nations in Africa are making efforts to protect streams and lakes that hold important stocks of tilapia. However, given the pressures on government resources, few nations can be expected to devote funds solely to aquatic ecosystems. A better approach is to combine any conservation efforts with those aiming to protect the more visible terrestrial species that are receiving so much attention. Governments are more likely to be able to expand an area being set aside as a park or nature reserve so as to include key aquatic systems than they are to establish a wholly separate area just for these poorly documented genetic resources.

As aquaculture holds great potential to contribute to future food supplies internationally, support for conservation efforts is needed vitally from governments outside the areas under threat. The principle of worldwide responsibility for what are essentially global resources that is being applied increasingly to such systems as tropical rain forests and the ozone layer holds equally well for aquatic genetic resources. Although we do not know yet just how much we stand to lose by not protecting aquatic habitats and species, nearly every loss is irreversible, and the world has one less tool in its efforts to provide food and livelihoods to its growing population.

[from Pullin (1990).]

ISA: International Study on Artemia

The International Study on Artemia is an informal, interdisciplinary group of researchers interested in brine shrimp, *Artemia* spp. It has been operative since 1976, and is headquartered at the Artemia Reference Centre in Ghent, Belgium. In contrast to most other species the preservation of brine shrimp is relatively simple. Their dry and dormant eggs, termed cysts, can be stored at ambient temperatures for up to 20 years. Hatching and raising young requires minimal equipment or technical expertise. Brine shrimp are used widely as live food for fish, especially those larval fish too small to accept prepared feeds.

The ISA maintains a collection of cysts collected from more than

100 geographical sites worldwide. As funding allows, ISA is characterizing its accessions and continuing to add to the collection, because it does not yet represent the available genetic diversity of *A. salina*.

STATUS OF SELECTED SPECIES

For a global overview of the management of aquatic genetic resources, responses to a set of questions were obtained in 1988 from nearly 40 scientists worldwide. They were contacted because of their knowledge of certain economically important aquatic animal species (see appendix) that were intensively managed and cultured or harvested (Table 4.3). Their responses, even after 7 years, provide an overview of the awareness and attitudes toward conservation and management of aquatic animal genetic resources.

The scientists were asked to provide four kinds of information on a particular species and its genetic conservation: if the species is included in a management programme; the extent of *in situ* or *ex situ* management; the level of knowledge about genetics and population structure; and the priority needs or principal constraints relating to conservation. Responses to the questions indicated that the management programmes, the human activities, and the genetic risks to which species are exposed vary (Tables 4.4 and 4.5).

Among the species surveyed, ecosystem maintenance is not a widely employed approach to managing aquatic genetic resources (Table 4.5). Although many national parks and sanctuaries include aquatic habitats and provide some degree of gene conservation, the parks were not established, nor are they managed, for that purpose.
One notable exception is Malawi Park and the associated Aquatic Zone, managed jointly by the Malawi Fisheries Department and the World Wide Fund for Nature for the purpose of *in situ* conservation of the habitat and indigenous cichlid species.

The harvesting regulations, culture practices, and other management activities related to these species can be considered species management. However, species management for the express purpose of conserving genetic resources was not identified for any of the species encompassed by the responses. Respondents did indicate that for most species genetics was at least considered when forming management plans and policies. The consensus of respondents was that contemporary natural resource management practices provide some degree of genetic resource conservation. All agreed that genetic conservation does not receive adequate priority in management programmes such as those dealing with modifications of environments, harvesting and culturing, or species introduction, transfer, and enhancement programmes.

At the national level, few countries include genetic conservation as a separate entity in their natural resource legislation, especially that relating to fisheries and aquaculture. However, most nations have enacted regulations to prevent overharvesting and to provide some control over pollution or other activities that alter aquatic habitats.

Table 4.3. Taxa included in responses to a questionnaire on aquatic genetic resources management.

	Family	Common Name	Scientific Name
Fishes	Chanidae	Milkfish	*Chanos chanos*
	Cichlidae	Tilapia	*Oreochromis* spp.
			Haplochromis spp.
	Coregonidae	Whitefishes, ciscoes	*Coregonus* spp.
			Prosopium spp.
			Stenodus spp.
	Cyprinidae	Common carp	*Cyprinus carpio*
		Asian carp	
		Grass carp	*Ctenopharyngodon idella*
		Silver carp	*Hypophthalmichthys molitrix*
		Bighead carp	*Aristichthys nobilis*
		Catla	*Catla catla*
		Rohu	*Labeo rohita*
		Mrigal	*Cirrhinus mrigala*
	Ictaluridae	N. American catfish	*Ictalurus* spp.
	Mugilidae	Mullet	*Mugil* spp.
	Salmonidae	Brook trout	*Salvelinus fontinalis*
		Lake trout	*Salvelinus namaycush*
		Arctic charr	*Salvelinus alpinus*
		Brown trout	*Salmo trutta*
		Atlantic salmon	*Salmo salar*
		Rainbow trout	*Oncorhynchus mykiss*
		Chum salmon	*Oncorhynchus keta*
		Pink salmon	*Oncorhynchus gorbuscha*
		Sockeye salmon	*Oncorhynchus nerka*
		Coho salmon	*Oncorhynchus kisutch*
		Chinook salmon	*Oncorhynchus tshawytscha*
		Cherry salmon	*Oncorhynchus masu*
Crustaceans	Phyllopodida	Brine shrimp	*Artemia salina*
	Penaeidae	Marine shrimp	*Penaeus* spp.
Molluscs	Ostreidae	Cupped oysters	*Crassostrea* spp.

Nations generally have legal controls on importing exotic or nonindigenous species for use in aquaculture or for introduction into open waters. Frequently the principal intent of such regulations is to inhibit the introduction of diseases and parasites. Invariably these regulations require all introductions or transfers to be accompanied by fish health certifications. The difficulty is with enforcement. Countries may impose restrictions centrally, or locally on farm sites, but usually they are difficult to enforce and often circumvented. A few countries still have no importation restrictions.

Table 4.4. Summary of responses to questionnaire about management activities for selected species.

	Management	Harvesting and Culturing			
Species Group	Modification of Environments	Capture Fisheries	Aquaculture	Culture Based Fisheries	Transfers and Intros
Tilapia	X	X	X		X
Common carp	X	X			
Bighead carp	X				
Catla		X			
Grass carp		X			
Mrigal		X			
Rohu		X			
Silver carp		X			
Milkfish		X	X		
Mullet		X	X		
North American catfish			X	X	
Arctic charr	X	X	X		
Brook charr	X	X	X	X	
Lake charr	X	X		X	X
Brown Trout	X	X	X	X	
Atlantic salmon	X	X	X	X	X
Rainbow trout		X	X	X	
Cherry salmon	X		X	X	
Chinook salmon	X	X	X	X	
Chum salmon		X			X
Coho salmon	X	X	X	X	
Pink salmon			X		X
Sockeye salmon		X			
Whitefishes, ciscoes	X	X		X	X
Cupped oysters	X	X			
Brine shrimp	X				
Marine shrimp	X	X			

Note: The management activities are organized according to the human interventions currently practised that may expose the species to genetic risks.

Several countries have outright bans on transfers in the interest of conserving their indigenous resources or protecting their native fisheries. Malawi is an exception among the African countries in its management of aquatic taxa and enforcement of protective legislation. Regulations on transfers of species in Malawi exist for the protection of indigenous populations, as well as restrictions controlling the exportation of certain cichlid taxa for ornamental fish trade. The country specifically bans the introduction of all exotic species to protect the unique fauna of Lake Malawi. Some states of India ban the introduction of all tilapia species as they are considered pests. The Philippines bans all exportation of milkfish fry, shrimp fry, and gravid shrimp spawners to protect its indigenous species from overexploitation. The United States and its territories ban the importation of several species, such as the green sea turtle (*Chelonia mydas*), that are

Table 4.5. Summary of responses to questionnaire about programmes for managing aquatic genetic resources.

	Management Programme			
Species Group	Ecosystem Maintenance	Species Management	Living Collection	Germplasm Storage
Tilapia	A	B	A, B, R	R
Common carp	B	B	A, B, R	R
Bighead carp	C	B	B, R	R
Catla	C	B	B, R	R
Mrigal	C	B	B, R	R
Rohu	C	B	B, R	R
Silver carp	C	B	B, R	R
Grass carp	C	B	B, R	R
Milkfish	C	B	R	C
Mullet	C	B	R	C
North American catfish	C	B	B, R	R
Arctic charr	B	B	B, R	C
Brook trout	B	B	B, R	C
Brown trout	B	B	B	R
Lake trout	B	B	B	C
Rainbow trout	B	B	B	R
Atlantic salmon	B	B	B	R
Cherry salmon	C	B	B	R
Chinook salmon	C	B	B	R
Chum salmon	C	B	B	R
Coho salmon	C	B	B	R
Pink salmon	C	B	B	R
Sockeye salmon	C	B	B	R
Whitefishes, ciscoes	C	B	C	C
Cupped oysters	C	B	B, R	B
Brine shrimp	C	B	A	A
Marine shrimp	C	B	B, R	B

Note: A = Programmes implemented for the express purpose of conserving genetic resources; B = management programmes that provide some degree of genetic resource conservation, but that were not implemented explicitly for that purpose; C = no conservation activity; and R = research programmes addressing conservation of genetic resources.

protected or endangered and the finished products, such as meat, skins, or shells, derived from them.

Management of each of the species addressed by respondents involves some form of artificial propagation in hatcheries. To the extent that hatchery stocks qualify as living collections, this form of *ex situ* management was reported for all the taxa. However, in all but three species, these collections were established and maintained for purposes other than genetic conservation. The exceptions were common carp in Hungary, African cichlids, and the brine shrimp collection of ISA. Establishing and maintaining living collections of cichlid species from Lake Victoria is being attempted at the Horniman Museum and at Chester Zoo (England) (Reid, 1994), and in the tilapia collection implemented by ICLARM. At Chester Zoo, of 136 species of fishes held in aquaria, 36 were bred in 1993-4, including 19 of 22 species classified by IUCN as conservationally sensitive (Reid, in press). The products of captive breeding at Chester are managed cooperatively with other institutions in Britain and elsewhere. For example, a shipment of *Tilapia guinasana*, an endangered species from sinkholes in Namibia, has been sent to the Desert Fishes Conservation Program in the USA, and several hundred juvenile *Haplochromis pyrrhocephalus*, a Lake Victoria cichlid, to the North American Lake Victoria Species Survival Program. This latter programme, directed from the New England Aquarium through the American Zoo and Aquarium Association, is intended to consolidate and focus support by the professional aquarium and academic communities on the conservation and ultimate restoration of representative remnants of the endemic fish fauna of the Lake Victoria Basin (Warmolts, 1994). Among its stated goals are the creation of experimental lakeside refugia, three to be established by the year 2000. Currently, it coordinates the activities of 24 participant institutions managing captive stocks of 32 species of haplochromines and *Oreochromis esculentus* in aquaria in North America. Molecular genetic studies of these captive stocks are being carried out at the Ohio State University, and cryopreservation programmes are being considered. There is a similar captive breeding programme for Lake Victoria fishes in mainland Europe coordinated from Artis Zoo, Amsterdam.

Since 1974 the Dexter National Fish Hatchery, Dexter, New Mexico, USA has maintained living collections of several endangered species of desert fishes native to the southwestern United States. Although the species are not commercially important, the collection may contain the only survivors of these species (Stuart & Johnson, 1981). Similar efforts have been made in the US Southwest by the nonprofit-making Desert Fishes Council (Office of Technology Assessment, 1985).

The responses indicated that the genetic history, population structure, and genetic diversity of managed populations have been or are being studied. Most studies are elucidating the structure of breeding populations as revealed by allelic variation using protein electrophoresis. The genetic history of some species has been studied in the context of postglacial distributions, but the recent history of most aquatic populations cannot be determined because the demographic and genetic information has not been collected. Although substantial effort has been expended to study the breeding organization of populations of some economically important species, little effort is directed toward understanding the genetic basis of other components of diversity, including morphological and morphometric variation, life history patterns, and ecological correlates.

All respondents agreed that education and research efforts must be increased to meet conservation goals. Professional and technical training programmes are needed to prepare scientists, managers, technicians, and teachers to implement genetic conservation programmes. It was argued that most contemporary university fisheries curricula do not provide adequate preparation in subjects that are prerequisite to understanding and practising conservation genetics. Continuing education opportunities for field biologists need to be expanded through technical training at the local level. Educational programmes are also required to expand public awareness of genetic conservation issue. Biological diversity and genetic conservation should become standard elements of primary and secondary curricula and part of the general education requirements of college and university students.

Respondents expressed the opinion that continuing research is required to provide essential information for managing aquatic genetic resources. The need for reliable methods of long-term germplasm storage for aquatic species, including cryopreservation of sperm, ova, and embryos, also was identified. Most other specific research needs identified were related to a species of interest. However, the unifying theme was that research is needed to develop practical methods for assessing the vulnerability of aquatic species to loss of their genetic resources, and for determining the effects of human activities on them.

RECOMMENDATIONS

Typically, conservation programmes are developed, implemented, and enforced at national, regional, and local levels. The effectiveness of national programmes often depends on international cooperation. Therefore the conservation of aquatic genetic resources clearly requires coordination of programmes at the highest levels of government. However, to be successful, programmes will need trained technical and professional staff and the support of an educated public.

National and global leadership is needed to coordinate activities to ensure the conservation of aquatic genetic resources.

Conservation of aquatic genetic resources requires coordinated global actions. For example, cooperation in the form of multilateral agreements will be needed to protect common or migratory resources. However, no international organization exists with the leadership, experience, and resources to coordinate, encourage, and implement conservation activities for aquatic animals. Present international efforts, such as those of the FAO, are important but cannot encompass the magnitude of the task to be accomplished. Nations must take a more active role in identifying their unique aquatic resources and developing efforts to protect and conserve them. For many nations unable to support conservation efforts, this will require multilateral or bilateral assistance.

The events that pose the greatest risks to aquatic environments and their fish and shellfish populations are complex and varied. Any single regional, national, or international organization is unlikely to have jurisdiction over all of them. A forum is needed for international dialogue about the industrial, agricultural, transportation, and other activities that affect aquatic resources negatively. One or more existing organizations may be able to serve effectively in this capacity. Conventions and treaties that seek to address the need to protect the world's oceanic resources must be broadly accepted and enforced.

Existing international fisheries organizations represent a potential infrastructure for fostering cooperation among nations in research and management of aquatic genetic resources. These organizations should work to establish international conventions, model policies for conservation, and guidelines for their implementation. They can be an important forum for review, revision, and establishment of international agreements and treaties that promote the conservation of aquatic genetic resources.

Education and training programmes need to be strengthened, and attention should be given to heightening public awareness of the importance of conserving aquatic genetic resources.

Many nations have research and educational programmes, in both the public and private sectors, that could be directed toward aquatic genetic conservation programmes. However, many of them are concerned with other aspects of marine and aquatic life and do not recognize fish and shellfish genetics as an important area of the aquatic sciences. Increasing public awareness of the importance of aquatic genetic resources should help to redirect research and educational capabilities toward programmes in genetic conservation of aquatic species, and to provide funding for these programmes.

Education is needed to increase public awareness of genetic resource conservation, how environmental degradation can contribute to the loss of biological diversity, and the latter's significance for future generations. The scale of effort must not be minimized. The education effort must demonstrate how actions taken on small segments of a problem can have cumulative effects in the long term, and must emphasize that conservation is a continuing activity. The education and training programmes must explore the policy limitations to conservation management fully, so that society develops an awareness of aquatic resource conservation. Finally, professional education and training programmes must be developed or improved to prepare resource managers, scientists, and technicians to manage and implement aquatic genetic conservation programmes and policies.

REFERENCES

Acere, T.O. 1988. The controversy over Nile perch, *Lates niloticus,* in Lake Victoria, East Africa. *Naga* (Oct.) 3-5.

Ahlstrom, E.H. & Radovich, J. 1970. Management of the Pacific sardine. In *A Century of Fisheries in North America* (Benson, N.G. ed.) *American Fisheries Society Special Report* **7**, 183-193.

Allen, R.L. & Goldsmith, M.D. 1981. Dolphin mortality in the eastern tropical Pacific incidental to purse seining for yellow fin tunas, 1979. *Report of the International Whaling Commission* **31**, 539-540.

Allendorf, F.W. & Leary, R.F. 1988. Conservation and distribution of genetic variation in a polytypic species, the cutthroat trout. *Conservation Biology* **2**, 170-184.

Allendorf, F.W., Ryman, N. & Utter, F.M. 1986. Genetics and fishery management: past, present, and future. In *Population Genetics and Fishery Management*, (Ryman, N. & Utter, F. eds.) Chap.1. Seattle: University of Washington Press.

Altukhov, Y.P. & Salmenkova, E.A. 1991. The genetic structure of salmonid populations *Aquaculture* **98**, 11-40.

Andrews, C. 1989. The size of the problem. *Fish* **1**, 1.

Anon. 1990. *Caring for the World: A Strategy for Sustainability.* Gland, Switzerland: International Union for the Conservation of Nature.

Anon. 1991a. The biology and conservation of rare fish. *Fisheries* **16 (1)**, 30.

Anon. 1991b. ICLARM: the world's leading aquaculture centre. *Biotechnology and Development Monitor* **7**, 12-13.

Anon. 1991c. *A strategic plan for international fisheries research. Part II (September 1991). ICLARM's response to the "TAC commentary on the draft strategic plan of ICLARM."* Manila: International Center for Living Aquatic Resources Management.

Bakke, T.A. 1991. A review of the inter- and intraspecific variability in salmonid hosts to laboratory infections with *Gyrodactylus salaris* Malmberg. *Aquaculture* **98**, 303-310.

Bakos, J. 1976. Crossbreeding Hungarian races of the common carp to develop more productive hybrids. In *Summary Report of the Technical Conference on Aquaculture, Kyoto, Japan, May 26, 1976.* FIR:AQ/Conf. **76/E.74.** Rome, Italy: Food and Agriculture Organization of the United Nations.

Bardach, J.E., Ryther, J.H. & McLarney, W.O. 1972. *Aquaculture: The farming and husbandry of freshwater and marine organisms.* New York: Wiley Interscience.

Barel, C.D.N., Dorit, R., Greenwood, P.H., Fryer, G., Hughes, N., Jackson, P.B.N., Kawanabe, H., Lowe-McConnell, R.H., Nagoshi, M., Ribbink, A.J., Trewavas, E., Witte, F. & and Yamaoka, K. 1985. Destruction of fishes in Africa's lakes. *Nature* **315**, 19-20.

Bartley, D.M. 1994. Towards increased implementation of the ICES/EIFAC codes of practice and manual of procedures for consideration of introductions and transfers of marine and freshwater organisms. *European Inland Fisheries Advisory Commission* **XVIII**/94/Inf.18.

Bergan, P.I., Gausen, D. & Hansen, L.-P. 1991. Attempts to reduce the impact of reared Atlantic salmon on wild in Norway. *Aquaculture* **98**, 319-324.

Briand, F. & Cohen, J.E. 1984. Community food webs have scale-invariant structure.

Nature **307**, 264-267.

Briand, F. & Cohen, J.E. 1987. Environmental correlates of food chain length. *Science* **238**, 956-960.

Buth, D. 1990. Genetic principles and the interpretation of electrophoretic data. In *Electrophoretic and Isoelectric Focusing Techniques in Fisheries Management.* (Whitmore, D.H. ed.) pp.1-21. Boca Raton: CRC Press.

Christy, F.T.Jr. & Scott, A. 1965. *The Common Wealth in Ocean Fisheries.* Baltimore, Md.: Johns Hopkins Press.

Cloud, J. & Thorgaard, G. eds. 1993. *Genetic Conservation of Salmonid Populations.* New York: Plenum Press.

Coe, J.M., Holts, D.B. & Butler, R.W. 1984. The "tuna-porpoise" problem: NMFS dolphin mortality reduction research, 1970-81. *Marine Fisheries Review* **46(3)**, 18-33.

Conrad, M. 1983. *Adaptability: the significance of variability from molecule to ecosystem.* New York: Plenum Press.

Consultative Group on International Agricultural Research (CGIAR). 1990. *International Centers Week 1990. Summary of Proceedings and Decisions.* Washington, D.C.: CGIAR Secretariat.

Doyle, R.W. 1983. An approach to the quantitative analysis of domestication selection in aquaculture. *Aquaculture* **33**, 167-185.

Ehrlich, P.R. & Ehrlich, A.H. 1981. *The Causes and Consequences of the Disappearance of Species.* New York: Random House.

FAO. 1981. Conservation of the genetic resources of fish: problems and recommendations. *FAO Fisheries Technical Paper* **217.** Rome, Italy: Food and Agriculture Organization of the United Nations.

FAO. 1993a. FAO Yearbook: Fishery Statistics; Catches and Landings, 1991. Vol. **72.** Rome, Italy: Food and Agriculture Organization of the United Nations.

FAO. 1993b. FAO Yearbook: Fishery Statistics; Commodities, 1991. Vol. **73**. Rome, Italy: Food and Agriculture Organization of the United Nations.

FAO. In press. Aquaculture Production 1986-1992. FAO Fisheries Circular 815 (Rev.6). Rome, Italy: Food and Agriculture Organization of the United Nations.

Ferguson, A. & Thorpe, J.E. eds. 1991. Biochemical genetics and taxonomy of fishes. *Journal of Fish Biology* **39 (Supplement A)**, 1-357.

Frankel, O.H. 1974. Genetic conservation: Our evolutionary responsibility. *Genetics* **78**, 53-65.

Froese, R. 1990. FISHBASE: an information system to support fisheries and aquaculture research. *Fishbyte* Dec., 21-24.

Gall, G.A.E. 1987. Inbreeding. In *Population Genetics and Fishery Management* (Ryman, N. & Utter, F. eds.) Chap.3. Seattle: University of Washington Press.

Gall, G.A.E., Bentley, B., Brodziak, J., Childs, E., Fox, S., Gomulkiewicz, R., Mangel, M., Panattoni, C. & and Qi, C. 1989. *Chinook mixed fishery project 1986-1989*. Davis: University of California. 30 pp.

Gjedrem, T. 1983. Genetic variation in quantitative traits and selective breeding in fish and shellfish. *Aquaculture* **33**, 51-72.

Gould, S.J. & Lewontin, R.C. 1979. The spandrels of San Marco and the Panglossian paradigm: a critique of the adaptationist programme. *Proceedings of the Royal Society of London* **B 205**, 581-598.

Grant, W.S., Milner, G.B., Krasnowski, P. & and Utter, F.M. 1980. Use of biochemical genetic variants for identification of sockeye salmon (*Oncorhynchus nerka*) stocks in Cook Inlet, Alaska. *Canadian Journal of Fisheries & Aquatic Sciences* **37**, 1236-1247.

Graves, J.E., Ferris, S.D. & Dizon, A.E. 1984. Close genetic similarity of Atlantic and Pacific skipjack tuna (*Katsuwonus pelamis*) demonstrated with restriction endonuclease analysis of mitochondrial DNA. *Marine Biology* **79**, 315-319.

Gulland, J.A. 1971. *The Fish Resources of the Oceans*. Farnham: Fishing News Books.

Hansen, L.P., Håstein, T., Nævdal, G., Saunders, R.L. & Thorpe, J.E. eds. 1991. Interactions between cultured and wild Atlantic salmon. *Aquaculture* **98**, 1-324.

Hartl, D. & Clark, A. 1989. *Principles of population genetics*. Sunderland: Sinauer Associates.

Hickling, C.F. 1962. *Fish Culture*. London: Faber and Faber.

Hilborn, R. 1985. Apparent stock recruitment relationships in mixed stock fisheries. *Canadian Journal of Fisheries & Aquatic Sciences* **42**, 718-723.

Hindar, K., Ryman, N. & Utter, F.M. 1991. Genetic effects of cultured fish on natural fish populations. *Canadian Journal of Fisheries & Aquatic Sciences* **48**, 945-957.

Huntingford, F.A. & Thorpe, J.E. 1992. Behavioural concepts in aquaculture. In *The Importance of Feeding Behavior for the Efficient Culture of Salmonid Fishes*. (Thorpe, J.E. & Huntingford, F.A. eds.) pp.1-4. *World Aquaculture Society, Workshops* **2**.

Hutchings, J.A. 1991. The threat of extinction to native populations experiencing spawning intrusions by cultured Atlantic salmon. *Aquaculture* **98,** 119-132.

International Council for the Exploration of the Sea. 1984. *Guidelines for Implementing the Code of Practice Concerning the Introduction and Transfer of Marine Species. Cooperative Research Report* **130.**

Jensen, K.W. & Snekvik, E. 1972. Low pH levels wipe out salmon and trout populations in southern Norway. *Ambio* **1**, 223-225.

Johnsen, B.O. & Jensen, A.J. 1991.The *Gyrodactylus* story in Norway. *Aquaculture* **98**, 289-302.

Jordan, W.C. & Youngson, A.F. 1991. Genetic protein variation and natural selection in Atlantic salmon (*Salmo salar* L.) parr. *Journal of Fish Biology* **39 (Supplement A)**, 185-192.

Joyner, T. 1980. Salmon ranching in South America. In *Salmon Ranching* (Thorpe, J.E. ed.) Chap.14. London: Academic Press.

Karr, J.R. 1981. Assessment of biotic integrity using fish communities. *Fisheries* **6(6)**, 21-27.

Kutkuhn, J.H. 1981. Stock definition as a necessary basis for cooperative management of Great Lakes fish resources. *Canadian Journal of Fisheries & Aquatic Sciences* **38**, 1476-1478.

Law, R. 1991. Fishing in evolutionary waters. *New Scientist*, March 2, 35-37.

Lewontin, R.C. 1984. Detecting population differences in quantitative characters as opposed to gene frequencies. *American Naturalist* **123**, 115-124.

Li, H.W., Schreck, C.B., Bond, C.E. & Rexstad, E. 1987. Factors influencing changes in fish assemblages of Pacific Northwest streams. In *Community and evolutionary ecology of North American stream fishes* (Matthews, W.J. & Heins, D.C. eds.) pp.193-202. Norman: University of Oklahoma Press.

Mathisen, O.A. 1989. Adaptation of the anchoveta *(Engraulis ringens)* to the Peruvian upwelling system. In *The Peruvian upwelling ecosystem: dynamics and interactions* (Pauly, D., Muck, P., Mendo, J. & Tsukuyama, I. eds.) pp.220-234. *ICLARM Conference Proceedings* **18**.

May, R.M. 1988. How many species are there on Earth? *Science* **241**, 1441-1449.

Miller, R.R., Williams, J.D. & Williams, J.E. 1989. Extinctions of North American fishes during the past century. *Fisheries* **14 (6)**, 22-38.

Moran, P., Pendas, A.M., Garcia-Vazquez, E. & Izquierdo, J.T. 1994. Electrophoretic assessment of the contribution of transplanted Scottish Atlantic salmon (*Salmo salar*) to the Esva River (Northern Spain). *Canadian Journal of Fisheries & Aquatic Sciences* **51**, 248-252.

Mork, J. 1991. One-generation effects of farmed fish immigration on the genetic differentiation of wild Atlantic salmon in Norway. *Aquaculture* **98**, 267-276.

Myers, N. 1988. Tropical forests and their species: Going, going ...? In *Biodiversity* (Wilson, E.O. ed.) pp.28-35. Washington, D.C.: National Academy Press.

NASCO (North Atlantic Salmon Conservation Organization). 1990. *Report on the Norwegian meeting on impacts of aquaculture on wild stocks.* Edinburgh: Paper **CNL(90)28**, 1-11.

Nehlsen, W., Williams, J.E. & Lichatowich, J.A. 1991. Pacific salmon at the crossroads: Stocks at risk from California, Oregon, Idaho, and Washington. *Fisheries* **16**, 4-21.

Nelson, K. & Soulé, M. 1987. Genetical conservation of exploited fishes. In *Population Genetics in Fishery Management* (Ryman, N. & Utter, F. eds.) pp.345-368. Seattle: University of Washington Press.

Nicolson, J. 1979. *Food From the Sea.* London: Cassell.

Office of Technology Assessment. 1985. *Grassroots Conservation of Biological Diversity in the United States.* Background Paper No. 1. OTA-BP-F-38. Washington, D.C.: US Government Printing Office.

Office of Technology Assessment. 1987. *Technologies to Maintain Biological Diversity.* OTA-F-330. Washington, D.C.: US Government Printing Office.

Paulik, G.J., Hourston, A.S. & Larkin, P.A. 1967. Exploitation of multiple stocks by a common fishery. *Journal of the Fisheries Research Board of Canada* **24**, 2527-2537.

Peters, J.C. 1982. Effects of river and streamflow alterations on fishery resources. *Fisheries* **7(2)**, 20-22.

Powers, D.A., Lauerman, T., Crawford, D., Smith, M., Gonzales-Villasenor, I. & DiMichele, L. 1991. The evolutionary significance of genetic variation at enzyme synthesizing loci in the teleost *Fundulus heteroclitus.* *Journal of Fish Biology* **39 (Supplement A)**, 169-184.

Pullin, R.S.V. 1988. *Tilapia Genetic Resources for Aquaculture.* Manila: International Center for Living Aquatic Resources Management.

Pullin, R.S.V. 1990. Down-to-earth thoughts on conserving aquatic genetic diversity. *Naga* January, 5-8.

Pullin, R.S.V. 1994. Biodiversity in Aquaculture. Paper presented to the International Forum on Biodiversity, 5-9 September 1994. Paris, France: UNESCO.

Ray, G.C. 1988. Ecological Diversity in Coastal Zones and Oceans. In *Biodiversity* (Wilson, E.O. ed.) pp.36-50. Washington, D.C.: National Academy Press.

Reeb, C.A. & Avise, J.C. 1990. A genetic discontinuity in a continuously distributed species: Mitochondrial DNA in the American oyster, *Crassostrea virginica. Genetics* **124**, 397-406.

Reid, G.McG. 1994. The conservation of small, genetically isolated popualtions of cichlid fishes. *Proceedings of the Conservation Genetics and Evolutionary Ecology Symposium*, Columbus, Ohio, November 1992.

Reid, G.McG. In press. Conservation programmes at Chester Zoo for fish and aquatic Invertebrates. *Memoire de l'Institut Océanographique Paul Ricard, Marseille.*

Ribbink, A.J. 1987. African lakes and their fishes: conservation scenarios and suggestions. *Environmental Biology of Fishes* **19**, 3-26.

Ricker, W.E. 1972. Hereditary and environmental factors affecting certain salmonid populations. In *The Stock Concept in Pacific Salmon* (Simon, R.C. & Larkin, P.A. eds.) pp.17-160. H.R.MacMillan Lectures in Fisheries. Vancouver, Canada: University of British Columbia.

Ricker, W.E. 1973. Two mechanisms that make it impossible to maintain peak-period yields from stocks of Pacific salmon and other fishes. *Journal of the Fisheries Research Board of Canada* **30**, 1275-1286.

Ricker, W.E. 1981. Changes in average size and average age of Pacific salmon. *Canadian Journal of Fisheries & Aquatic Sciences* **38**, 1636-1656.

Riddell, B.E. 1986. Assessment of selective fishing on the age at maturity in Atlantic salmon (*Salmo salar*): A genetic perspective. *Canadian Special Publication in Fisheries and Aquatic Sciences* **89**, 102-109.

Roberts, L. 1990. Zebra mussel invasion threatens U.S.waters. *Science* **249**,1370-1372.

Ryman, N. 1983. Patterns of distribution of biochemical genetic variation in salmonids: differences between species. *Aquaculture* **33**, 1-21.

Ryman, N. 1991. Conservation genetics and considerations in fisheries management. *Journal of Fish Biology* **39 (Supplement A)**, 211-224.

Ryman, N. & Laikre, L. 1991. Effects of supportive breeding on the genetically effective population size. *Conservation Biology* **5**, 325-329.

Ryman, N. & Utter, F. eds. 1986. *Population genetics and fishery management.* Seattle: University of Washington Press.

Saunders, R.L. 1991. Potential interaction between cultured and wild Atlantic salmon. *Aquaculture* **98**, 51-60.

Schom, C.B.& Davidson, L.A. 1982. Genetic control of pH resistance in Atlantic salmon *(Salmo salar). Canadian Journal of Genetics & Cytology* **24,** 636.

Schonewald-Cox, C.M., Chambers, S.M., MacBryde, B. & Thomas, W.L. 1983. *Genetics and Conservation.* Menlo Park, Calif.: Benjamin-Cummings.

Shaklee, J.B., Klaybor, D.B., Young, S. & White, B.A. 1991. Genetic stock structure of odd-year pink salmon, *Oncorhynchus gorbuscha* (Walbaum), from Washington and British Columbia and potential mixed stock fisheries applications. *Journal of Fish Biology* **39 (Supplement A)**, 21-34.

Shelbourne, J.E. 1971. *Artificial Propagation of Marine Fish.* Jersey City, N.J.: TFH.

Simon, R.C., McIntyre, J.D. & Hemmingsen, A.R. 1986. Family size and effective population size in a hatchery stock of coho salmon (*Oncorhynchus kisutch*). *Canadian Journal of Fisheries & Aquatic Sciences* **43**, 2434-2442.

Smith, B.R. 1971. Sea lampreys in the Great Lakes of North America. In *The Biology of Lampreys* (Hardisty, M.W. & Potter, eds.) Vol.I, chap.5. London: Academic Press.

Smith, P.J., Francis, R.I.C.C. & McVeagh, M. 1991. Loss of genetic diversity due to fishing pressure. Fisheries Research **10**, 309-316.

Soulé, M.E., ed. 1986. *Conservation biology: the science of scarcity and diversity.* Sunderland, Mass.: Sinauer Associates.

Soulé, M.E. & Wilcox, B.A. 1980. *Conservation biology: an evolutionary-ecological perspective.* Sunderland, Mass.: Sinauer Associates.

Spencer, C.N., McClelland, B.R. & Stanford, J.A. 1991. Shrimp stocking, salmon collapse, and eagle displacement. *BioScience* **41**, 14-21.

Stock Concept International Symposium. 1981. Proceedings of the 1980 Stock Concept International Symposium. *Canadian Journal of Fisheries & Aquatic Sciences* **38**, 1457-1921.

Stoss, J. 1983. Fish gamete preservation and spermatozoan physiology. In *Fish Physiology* (Hoar, W.S., Randall, D.J. & Donaldson, E.M.eds.) Vol. IX, Part B, 305-350. New York: Academic Press.

Stuart, H. & Johnson, J.E. 1981. A refuge for southwestern fish. *New Mexico Wildlife* **26**, 2-5.

Tave, D. 1986. *Genetics for fish hatchery managers.* Westport, Conn.: AVI.

Thompson, N.B. 1988. The status of loggerhead, *Caretta caretta*, Kemp's ridley *Lepidochelys kempi*, and green, *Chelonia mydas*, sea turtles in U.S. waters. *Marine Fisheries Review* **50(3)**, 16-23.

Thorne-Miller, B. & Catena, J.G. 1991. *The living ocean. Understanding and protecting marine biodiversity.* Washington, D.C.: Island Press.

Thorpe, J.E. ed. 1980. *Salmon Ranching*. London: Academic Press.

Thorpe, J.E. 1991. Acceleration and deceleration effects of hatchery rearing on salmonid development, and their consequences for wild stocks. *Aquaculture* **98**, 111-118.

Thorpe, J.E. 1993. Impacts of fishing on genetic structure of salmonid populations. In *Genetic Conservation of Salmonid Populations*. (Cloud, J. & Thorgaard, G. eds.) pp.68-81. New York: Plenum Press.

Thorpe, J. E. & Huntingford, F.A. eds. 1992. *The importance of feeding behavior for the efficient culture of salmonid fishes*. *World Aquaculture Society Workshops* **2**.

Thorpe, J.E. & Mitchell, K.A. 1981. Stocks of Atlantic salmon (*Salmo salar*) in Britain and Ireland: discreteness and current management. *Canadian Journal of Fisheries & Aquatic Sciences* **38**, 1576-1590.

Turner, H.J. 1977. Changes in size structure of cichlid populations of Lake Malawi resulting from bottom trawling. *Journal of the Fisheries Research Board of Canada* **34**, 232-238.

Utter, F.M. 1991. Biochemical genetics and fishery management: an historical perspective. *Journal of Fish Biology* **39 (Supplement A)**, 1-20.

Utter, F.W., Campton, D., Grant, S., Milner, G. & Seeb, J. 1980. Population structures of indigenous salmonid species of the Pacific northwest. In *Salmonid Ecosystems of the North Pacific* (McNeil, W.J. & Himsworth, D.C. eds.) pp.285-304. Corvallis: Oregon State University Press.

Utter, F., Aebersold, P. & Williams, G. 1986. Interpreting genetic variation detected by electrophoresis. In *Population Genetics and Fishery Management* (Ryman, N. & Utter, F. eds.) pp.21-45. Seattle: University of Washington Press.

Waples. R.S. 1987. A multispecies approach to the analysis of gene flow in marine shore fishes. *Evolution* **41**, 385-400.

Waples, R.S. 1991. Conservation genetics of Pacific salmon. II. Effective population size and the rate of loss of genetic variability. *Journal of Heredity* **81**, 267-276.

Waples, R.S., Winans, G.A., Utter, F.M. & Mahnken, C. 1990. Genetic approaches to the management of Pacific salmon. *Fisheries* **15**, 19-25.

Ward, R.D., Woodwark, M. & Skibinski, D.O.F. 1994. A comparison of genetic diversity levels in marine, freshwater and diadromous fishes. *Journal of Fish Biology* **44**, 213-232.

Warmolts, D.I. 1994. Conservation and the Lake Victoria Basin: the role of North American zoos and aquariums. *Aquatic Survival* **3 (2)**, 1, 7-11.

Warren, C.E. 1971. *Biology and Water Pollution Control.* Philadelphia: W.B.Saunders.

Welcomme, R.L. 1988. International transfers of inland aquatic species. In *FAO Fisheries Technical Paper* **294**. Rome, Italy: Food and Agriculture Organization of the United Nations.

Wheeler, A. & Sutcliffe, D. 1990. The biological conservation of rare fishes. *Journal of Biology* **37 (Supplement A)**, 1-271.

Wilson, E.O., ed. 1988. *Biodiversity.* Washington, D.C.: National Academy Press.

Wolf, E.C. 1988. Avoiding a mass extinction of species. In *State of the World 1988* . (Starke, L. ed.) pp.110-117. New York: W.W.Norton & Company.

Zalewski, M., Thorpe, J.E. & Gaudin, P. eds. 1991. *Fish and Land/InlandWater Ecotones.* Lodz: UNESCO/MAB, University of Lodz.

APPENDIX

The following is a questionnaire sent to about 40 scientists worldwide, having knowledge of various economically important fish and shellfish species (see Chapter 4).

Respondent: Please provide your views on the following questions related to [species]:

I. Description of resource(s):
   A: What species and populations are being managed?
   B: What is the purpose of the management programme(s)?
      (e.g., conservation, food production, recreation, aesthetics, other)
   C: What management approaches are employed?
      (e.g., approach to conservation; approach to food production, such as capture fisheries, culture-based fisheries, aquaculture)
   D: What administrative and other activities promote the goal?
      (e.g., education, research, information, public policy, regulation, enforcement)

II: Genetic conservation:
   A: Is genetic conservation included in the management programme? If so, what are the genetic conservation activities?
      On-site management:
         a: Ecosystem maintenance (e.g., national parks)
         b: Species management (e.g., refuges)
      Off-site management:
         a: Living collections (e.g., captive breeding programmes)
         b: Germplasm storage (e.g., semen, ova or embryo banks)
   B: What is known about the genetic history, population structure, and genetic diversity of the managed populations?
   C: What are the constraints to genetic conservation?
      (e.g., scientific or technical, social or political, legal, economic)
   D: How far has genetic conservation proceeded? Can specific accomplishments be cited?
   E: What are the principal needs in genetic conservation?
      (e.g., methods, information including research or extension, policy, enforcement, personnel, facilities, funding)

III. Please provide any other information that you feel is relevant.

PART TWO
MANAGEMENT EXAMPLES

5
THE ATLANTIC SALMON

J.E.Thorpe & L.Stradmeyer

Salmon are opportunistic, phenotypically plastic generalists, that have highly variable life history strategies within species, notably through variable developmental rates (Thorpe, 1986). Two major symposia reviewed knowledge of their heritable characteristics (Simon & Larkin, 1972; Stock Concept International Symposium, 1981). In each case, participants presented evidence demonstrating the genetically discrete stock structure characteristic of these fishes. The later meeting made some attempts to evaluate and predict the nature and extent of human impacts on the genetic structure of some of these species. More recently, Ståhl (1987) reviewed information on the population genetics of Atlantic salmon (*Salmo salar*), and noted the existence of at least three distinct groups: North American, eastern Atlantic, and Baltic. He also recorded evidence of genetic instability and reduced variation among hatchery populations (see also Verspoor, 1988), which he attributed to the use of too few parental fish.

The purpose of this management study is to describe what is known of the extent of variation in Atlantic salmon, to consider the effect of major human impacts on this variety, and to record present approaches to its genetic conservation.

STOCK STRUCTURE IN *SALMO SALAR*

Nineteenth century taxonomists divided salmonid species into many different subtaxa on account of their varied appearance, but particularly because of their range of life history strategies. For example, some landlocked forms of Atlantic salmon in North America were given subspecific status as *S. salar sebago*. However, recent protein polymorphism (Ståhl, 1987) and mitochondrial DNA analyses (Birt *et al.*, 1986) do not endorse such a separation in this species.

Evidence for distinct genetic variation within the species has been accumulating rapidly during the past 30 years. Cytological studies have found variation in the species' chromosome number. Boothroyd (1959) postulated that the diploid number of 56 in North America had evolved through two centric fusions or misdivisions from that of 60 in European stocks. However, researchers have found diploid numbers varying from 55 to 60 in various European populations (Rees, 1967; Nygren *et al.*, 1972; Grammeltvedt, 1975; Gjedrem *et al.*, 1977; Zelinskiy & Medvedeva, 1985). The picture is complicated further by

Table 5.1. Biochemical genetic variation in Atlantic salmon: gene diversity analyses of Atlantic salmon stocks (after Ståhl, 1987).

Item	Hatchery Stocks	Wild Stocks	Total
Number of:			
Regions*	4	3	4
Drainages	17	18	31
Samples	24	29	53
Fish	2410	1699	4109
Absolute gene diversity			
Total	0.037	0.041	0.040
Standard error	0.020	0.021	0.021
Relative gene diversity (%)			
Between regions	26.5	29.7	28.4
Between drainages within regions	14.5	4.9	9.0
Between samples within drainages	5.1	1.5	3.6
Within samples	53.9	63.9	59.0

* The Baltic Sea, eastern and western Atlantic Ocean, and landlocked populations.

chromosome polymorphism within individuals (Barsiene, 1981; Hartley, 1987). The species differs from all other salmonids in having a chromosome arm number (NF) of 72 or 74. Differences in this number have been found between fish from Europe (NF = 74: Rees, 1967) and North America (NF = 72: Roberts, 1970), although Barsiene (1981) reported both arm numbers in fish from Russian populations.

Electrophoretic analyses have demonstrated decisively that the species is composed of separate genetically discrete stocks (Nyman, 1967; McKenzie & Paim, 1969; Møller, 1970; Payne, 1980; Slynko *et al.*, 1981; MacCrimmon & Claytor, 1986; Altukhov & Salmenkova, 1987; Ståhl, 1987). Ihssen *et al.* (1981) defined a stock as an intraspecific group of randomly mating individuals with temporal or spatial integrity. Ståhl (1987) analyzed data on allele frequencies at 38 loci, 9 of which were polymorphic, from 45 European and 8 North American samples (Tables 5.1 and 5.2). In wild populations he found that 63.9% of the total gene diversity was within stocks. He noted that 29.7% was found between salmon from the three main regions: Baltic, eastern Atlantic, Western Atlantic and landlocked stocks; 4.9% was due to differences between river stocks within regions; and 1.5% was due to differences between populations within rivers. Davidson *et al.*, (1989a) suggested that this implied that either homing is not as accurate as is assumed, or that the present populations have been derived very recently from a common population. This latter explanation is more likely, as Atlantic salmon have colonized their present range since the last glaciation, over a period of 6000-15,000 years. Only 1 salmon per generation need stray from one stock to another and breed successfully to prevent different neutral alleles at the same locus from being fixed in the different populations (Slatkin, 1987).

Table 5.2. Distribution of gene diversity at 11 variable loci among the 53 samples of Table 5.1.(Averages based on 38 loci.) (after Ståhl, 1987).

	Absolute gene diversity		Relative gene diversity (%)			
	Total	Within samples	Between regions	Between drainages within regions	Between samples within regions	Within samples
*AAT-3**	0.4100	0.2030	39.70	8.3	2.4	49.6
*AGP-2**	0.0009	0.0008	0.04	1.3	-	98.7
*IDH-3**	0.0060	0.0050	0.50	16.0	-	83.5
*LDH-4**	0.0009	0.0009	0.30	0.5	0.9	98.3
*MEP-2**	0.4970	0.2560	35.20	11.7	1.5	51.6
*MDH-1**	0.0100	0.0090	1.30	2.2	1.7	94.7
*MDH-3**	0.0760	0.0590	17.00	3.0	2.3	77.7
*PGI-1**	0.0010	0.0010	0.10	1.2	0.2	98.6
*PGM-1**	0.0080	0.0070	0.80	4.1	5.9	89.2
*SDH-1**	0.4920	0.3430	15.50	7.9	6.9	69.7
*SDH-2**	0.0020	0.0020	0.10	4.6	-	95.3
Average	0.0400	0.0230	28.40	9.0	3.6	59.0
Standard error	0.0210	0.0120	5.80	1.1	1.3	5.2

Ståhl (1987) noted that the greatest regional difference lay between the stocks of North America and Europe. Early studies differentiated North American from European populations on the grounds of transferrin allele frequencies, and subsequent ones using restriction analyses of mitochondrial DNA (Bermingham *et al.*, 1991), ribosomal RNA genes (Cutler *et al.*, 1991) and direct sequence analysis of the cytochrome b gene (McVeigh *et al.*, 1991), have confirmed this continental difference.

Ståhl (1987) found that the landlocked forms were closely similar to neighbouring anadromous ones, implying derivation of the former from the latter. Vuorinen & Berg (1989) showed genetic divergence between anadromous and resident salmon in the River Namsen, Norway. Although no fixed allele frequency difference was found, 18.3% of the genetic diversity occurred between types, and Nei's (1973) distance between the populations was 0.01, which is relatively high for Atlantic salmon populations. Mean heterozygosity of the anadromous fish was 3.1%, while that of the residents was only 1.0%. The authors suggested that this latter low value probably represents a founder effect, due to the relatively few females that may have matured originally as parr when the resident population formed in response to the formation of the Trongfoss waterfall about 9500 years ago. Sutterlin & MacLean (1984) found developmental differences between the progeny of anadromous and non-anadromous salmon in Newfoundland, when these were reared under the same environmental conditions, and Birt *et al.*, (1991a) confirmed this when they compared the progeny of the two types living sympatrically. Differences in developmental timing may ensure the maintenance of genetic distinction in sympatric populations, and may account for the persistence of this distinction in the River Namsen despite the annual stocking for 27 years of 20,000-100,000 progeny of anadromous salmon into the habitat of the resident population. The genetic distinctness of sympatric populations of resident and anadromous salmon in Newfoundland was established by protein electrophoresis (Verspoor & Cole, 1989), and by mitochondrial DNA analysis (Birt *et al.*, 1991b).

Ryman (1983) highlighted the remarkably high fraction of diversity between stocks (at 21.4% in his study, and 41% in Ståhl's, two to four times that found between the major racial groups of humankind). Ståhl (1987) also noted that the relative genetic divergence between samples was about three times as high for hatchery stocks within and between stocks in regions as for wild stocks. He attributed this to greater genetic drift when stocks are propagated artificially.

The amount of genetic variation in Atlantic salmon is low compared with freshwater fish in general (Davidson *et al.*, 1989a), and is more similar to that found in marine species (Gyllensten, 1985). The relatively low genetic diversities of Atlantic salmon populations of about 4% (Table 1), when compared with those of other teleost fishes (e.g., 7.8% average for 14 species according to Selander [1976]), have been attributed to their tetraploid origin (Ohno *et al.*, 1968), the duplicated loci being assumed to reduce the need for polymorphism. Protein electrophoretic surveys showed that 4 gene loci (*AAT**, *MDH**, *ME**, and *SDH**) accounted for 95% of the total detectable variation (Davidson *et al.*, 1989a). Cross & Ward (1980) have also suggested that the relatively small size of breeding populations leads to these low levels of variation, compared, for example, with those of 9% found in the large panmictic populations of plaice (*Pleuronectes platessa*) (Ward & Beardmore, 1977).

Ståhl's (1987) studies using multiple enzyme polymorphisms indicated differences at a range of river system and tributary levels, and recently Koljonen (1989) reported differences on a similar geographic scale between Finnish and Russian populations. Such regional differences imply relatively long periods of reproductive isolation, which are usually attributed to physical subdivision of ancestral populations as a consequence of glaciation. From allele frequency data at 6 polymorphic loci, from salmon taken in rivers draining into the Barents, White, and Baltic seas, Kazakov & Titov (1991) suggested that Atlantic salmon colonized the Onega and Pechora river basins from the Baltic drainage as the last ice-sheet receded between 6000 and 11,000 years ago, but that the rivers in the Kola Peninsula were colonized separately from the Barents Sea.

Within continents, examinations of many isozyme systems have confirmed that not only are the populations of different river systems different (Khanna *et al.*, 1975; Cross & Payne, 1977; Cross *et al.*, 1978; Cross & Healy, 1983; Verspoor, 1986; Ståhl, 1987; McElligott & Cross, 1991), but that at least some rivers contain more than one distinct stock (Ståhl, 1987; Verspoor & Cole, 1989; McElligott & Cross, 1991; Verspoor *et al.*, 1991). The interpretation of such differences is complicated by year-to-year fluctuation in allele frequencies, due to varying proportions of a year-class contributing to spawning in any one year, and to the overlapping of year-classes in spawning populations (Waples & Teel, 1990).

At a still finer scale of discrimination, restriction endonuclease analysis of salmon mitochondrial DNA (Gyllensten & Wilson, 1987; Davidson *et al.,* 1989b; Hovey *et al.*, 1989; Palva *et al.*, 1989; Birt *et al.*, 1991b) has confirmed the locally restricted nature of breeding groups, but much remains to be done to define their actual physical extent (Ståhl, 1983). The implication is that *Salmo salar* exists as a species formed of thousands of reproductively isolated populations (Power, 1981; Saunders, 1981; Thorpe & Mitchell, 1981).

Morphological and developmental variation are less dependable sources of information on inherent variety within the species, since such characteristics as size, shape, coloration, and age at migration and at maturity are the expressions of genetic action conditioned by environmental opportunity. However, the relative constancy of differences between the early rearing environments of different stocks is reflected in potentially diagnostic structural and developmental differences between these stocks. The patterns of scale growth vary sufficiently between regions to have proved useful as a discriminatory characteristic of broad geographic groups, as, for example, in the mixed stock fishery off the Greenland coast (Jensen & Lear, 1980; Reddin & Misra, 1985; Reddin & Short, 1985). Comparative analyses of scale characteristics and serum electrophoresis in fish sampled off Greenland have yielded similar estimates of the proportions attributable to the two continents of origin (Lear & Sandeman, 1980).

Reddin & Misra (1985) used four scale characteristics to identify fish of a single river stock from a mixed stock fishery, using Hotelling's T^2 multivariate statistical test. Also from scale pattern data, Thorpe & Mitchell (1981) showed that specific growth rates of fish from separate Scottish river populations differed during their first year at sea, again implying differences in their pattern of usage of the marine environment. Scale patterns of freshwater growth may also be useful to distinguish wild salmon from those reared in hatcheries (Antere & Ikonen, 1983) and to these distinctions Hansen *et al*. (1987a) added morphometric data. Investigators have successfully used discriminant analysis using morphometric character sets (standard length, paired fin lengths, body depth, gape width) to distinguish separate river stocks within Newfoundland, Scotland, and Sweden (MacCrimmon & Claytor, 1985, 1986).

In different tributaries of the Miramichi River, Canada, Riddell *et al*., (1981) found morphologically separable populations distinguished by differences in body shape and fin proportions. They established the genetic nature of these differences in laboratory rearing experiments, and considered them to be adaptively related to differences in hydrological conditions. Subsequently, within these Miramichi tributaries, Ståhl *et al*., (1983) found genetically different local stocks, and implied that these genetic differences might be coupled with the morphological ones Riddell *et al*., (1981) described. In the Alta River, Norway, Heggberget *et al*., (1986) determined local genetic differences between populations in three, consecutive 15-kilometre stretches of the main river. These differences were expressed both at electrophoretically detectable protein loci, and in consistently different growth and smolting patterns.

Biological tags, such as the nematode parasite *Anisakis simplex*, have provided corroborative evidence of distinctions between regional populations. The density of larvae in Scottish salmon was more than twice that in English and Irish fishes (Beverley-Burton & Pippy, 1978), suggesting that salmon from these separate nursery areas feed in different regions at sea. Differences in the frequencies of acid phosphatase phenotypes of the parasites in the salmon of different geographic origin (Beverley-Burton, 1978) also support this hypothesis.

Discontinuity of genetic variation within Atlantic salmon and apparent stability within stocks imply that Atlantic salmon have efficient behavioural and physiological mechanisms for maintaining their discreteness. Precise homing of adults to their natal river at spawning is well documented (Behnke, 1972; Saunders, 1981; Thorpe & Mitchell, 1981; Stabell, 1984; Hansen *et al*., 1987b), and homing to specific locations on that river is now implied (Couturier *et al*., 1986; Heggberget *et al*., 1986). Measured straying rates are low, usually about 2-3% (Went, 1969; Larsson, 1974; Hawkins *et al*., 1979), but Hansen & Lea (1982) recorded values up to 18% in some Norwegian systems. However, from isozyme analyses Ståhl (1981) concluded that the genetic migration rate between several Swedish stocks was less than one individual per year; much lower than that expected from a numerical straying rate of 50 to 200.

Since salmonids are famous for their homing precision, the existence of straying is generally overlooked, and when found tends to be regarded as a failure by the straying individuals to achieve the population norm. However, without straying there would be no salmonid populations throughout much of their present range in the Atlantic drainages, as much of that area has been colonized by salmon since the last glaciation - a matter of only 6000-15,000 years. Straying is a normal population strategy entirely appropriate to a colonising phase in a salmon population (Thorpe, 1994d), as evidenced from the spread of introduced chinook salmon (*Oncorhynchus tschawytsha*) in New Zealand (McDowall, 1990), the successful establishment of various *Oncorhynchus* species in the Laurentian Great Lakes (Kwain & Lawrie, 1981), and the current invasion of new habitats by Dolly Varden (*Salvelinus malma*) and sockeye salmon as glaciers retreat in Alaska (Milner, 1987; Milner & Bailey, 1989). The apparent absence of large-scale straying between established salmonid populations implies a selective advantage for precise homing when all available habitats are occupied, and further implies the existence of barriers to effective interbreeding between these particular stocks, but the nature of those barriers is unknown. The inference is that there are adaptive genetic differences between stocks, and that the frequent failures of deliberate transplantation programmes may be attributable to this need for close matching of stock and habitat (Saunders 1981).

DEVELOPMENTAL CHARACTERISTICS

The developmental characteristics of salmon discussed below are within the areas of salmonid growth, smolting, maturation, and return timing and rate.

Salmonid growth

The plasticity of salmonid growth makes determining the extent of genetic and environmental influences difficult. Investigators have found considerable variations in growth rates between stocks of Atlantic salmon and between sib families within a stock (Thorpe, 1977; Naevdal *et al.*, 1978; Refstie & Steine, 1978). Under simplified and controlled conditions much of this variance can be ascribed to differences of genetic origin (Thorpe, 1977; Gunnes & Gjedrem, 1978; Thorpe *et al.*, 1983), but environmental components such as tank effects and food supply should still be considered (Ricker, 1972; Thorpe, 1977, 1987a; Refstie & Steine, 1978).

Circumstantial evidence from wild stocks suggests differences in growth at sea (Thorpe & Mitchell, 1981), but this could be attributable to local environmental differences in the coastal areas the migrant juveniles first encounter (Scarnecchia, 1983). Further, as differences in the pattern of commercial catches between groups of stocks within Scotland suggested differences in use of the marine environment (Thorpe & Mitchell, 1981), growth differences between these stocks could be entirely environmentally induced.

Under controlled hatchery conditions, researchers have found differences in heritability estimates for growth when calculated separately for dam and sire components. Several studies found a higher dam component, suggesting the presence of maternal effects or nonadditive genetic variation (Refstie *et al.*, 1977; Thorpe, 1977; Gunnes & Gjedrem, 1978; Refstie & Steine, 1978; Friars *et al.*, 1979; Gjerde & Gjedrem, 1984). Naevdal *et al.* (1975) found similar heritability for weight but higher estimates for length based on a dam than a sire component. As some of these experiments practised culling to reduce population size, which may not have been random with respect to body size (Gjerde, 1984a), some of the findings should be treated with caution. Heritability estimates depend on the conditions under which they are obtained, so when considering the variance of growth in wild populations, their value lies in demonstrating the reality of fundamental genetic control of the growth rate.

Thorpe & Morgan (1978) showed that the male parent had a greater influence on the rate of development than the female parent, while the latter had a more pronounced influence over absolute body size during the first year of growth. Glebe *et al.* (1979) investigated this maternal effect and suggested it was a consequence of differences in egg size.

Investigators have determined that growth arrest after July, consequent on reduced food intake (Higgins, 1985) due to reduced feeding motivation (Metcalfe *et al.*, 1986), is characteristic of components of sibling juvenile salmon populations in freshwater (Thorpe, 1977, 1987a; Bailey *et al.*, 1980; Baglinière & Maisse, 1985; Kristinsson *et al.*, 1985). This leads to a bimodal segregation of faster and slower growth groups in the autumn. This growth segregation has now been shown to occur in wild populations also (Baglinière & Maisse, 1985; Baglinière & Champigneulle, 1986; Nicieza *et al.*, 1991; Huntingford *et al.*, 1992). Both parental origins and environmental variables affect variation in the proportions of populations in the different growth modes (Bailey *et al.*, 1980; Thorpe, 1987a) and in the timing of their segregation (Thorpe, 1986, 1987a; Villarreal *et al.*, 1988). In a unified model of growth, smolting, and maturation, Thorpe (1986) postulated that the physiological decision to maintain or arrest growth was taken during a limited interval in late summer, and that the direction of that choice depended on a regulator of appetite, genetically determined through a threshold of developmental performance related to growth opportunity at that time. Since the fish feed primarily by day (Higgins & Talbot, 1985), Thorpe *et al.* (1989) showed experimentally that an index of this growth opportunity was given by the product of daily mean temperature and hours of daylight (the thermal sum).

A reasonable expectation is that food intake is correlated with social dominance (Metcalfe *et al.*, 1989), and Rosenau & McPhail

(1987) have demonstrated a stock-specific difference in levels of agonistic behaviour in juvenile coho salmon (*Oncorhynchus kisutch*). Similar differences are likely to be found in Atlantic salmon, but information on this is not yet available.

Sadler *et al.* (1986) found that food conversion efficiency differed between stocks at two different temperatures, being more efficient the closer the temperature was to their preferred temperature. They also showed stock differences in growth rates at different temperatures. Likewise Gunnes (1979) found a lower optimum temperature for alevin development among Norwegian salmon than Peterson *et al.* (1977) did among Canadian salmon. Gunnes suggested that Norwegian stocks were adapted to colder regimes than Canadian stocks. Jordan & Youngson (1991) found differential growth among *MEP-2** genotypes of parr, and suggested a connection here with the established correlation between temperature and *MEP-2** allele frequencies (Verspoor & Jordan, 1989). Torrissen (1987, 1991) and Torrissen *et al.*, (1993) found higher growth rate among salmon possessing the trypsin isozyme *TRP-2(92)* than among those with the other variants at this locus, during both the first feeding period and again between smolting and maturation. Feed conversion was more effective, and protein efficiency and specific growth rate were higher in the *TRP-2(92)* fish (Torrissen & Shearer, 1992).

Smolting

The modal size group that maintains appetite in late summer and continues growth through the winter smolts the following spring, whereas the group whose appetite falls off does not. Hence age at smolting is also dependent on growth opportunity, being inversely correlated with the rate of growth (Thorpe, 1986). As the nature and regulation of all the component processes of smolting are incompletely understood (Bern & Mahnken, 1982; Thorpe *et al.*, 1985; Hansen *et al.*, 1989a; Saunders *et al.*, 1994), the environmental and genetic regulators of age at smolting are currently inseparable from those directly influencing growth. In wild populations, researchers have noted that age at smolting increases with latitude and decreases with mean temperature and length of growing season (Power, 1981). At a given latitude North American salmon appear to smolt at older ages than those in Europe. However, using data from 182 river populations, Metcalfe & Thorpe (1990) found that this difference is attributable principally to differences in thermal sum properties of the salmon's habitats. Jensen & Johnsen (1986) considered that some salmon stocks were physiologically adapted to grow at unusually low temperatures, but critical experiments were lacking.

This dominance of environmental effects on age at smolting is further demonstrated by the commonplace observation that improved growth opportunities in hatcheries result routinely in higher growth rates and lower ages at smolting than in the wild (Thorpe, 1991). However, hatchery experiments have also revealed significant stock (Refstie *et al.*, 1977) and family differences (Thorpe, 1977, 1987a) in age at smolting, with highly significant sire x dam interactions suggesting considerable nonadditive genetic variance (Refstie *et al.*, 1977).

Riddell & Leggett (1981) found marked differences in the timing of downstream migration of juvenile fish in the Sabbies River and Rocky Brook, New Brunswick, Canada. In breeding experiments they established evidence for the separate stock identity of these two populations, but were unable to show that the timing of migration was itself genetically determined. However, Struthers & Stewart (1986) showed that a population transplanted from a downstream to an upstream locality on the River Tay system, Scotland, migrated consistently about one month later than the native stock during four successive years.

Hurley & Schom (1984) found heritable differences in swimming stamina in juvenile salmon, but did not relate these to differences in habitat.

As smolting implies a loss of freshwater adaptations and maturation requires their retention (Thorpe, 1987b), these developmental processes are at least partially conflicting. The existence of wholly freshwater populations of salmon shows that smolting is not essential for the completion of the life cycle, and so should be considered physiologically subordinate to maturation (Thorpe, 1994a). Hence age at smolting is influenced by the opportunity for parr maturation (Bailey *et al.*, 1980; Saunders *et al.*, 1982; Baglinière & Maisse, 1985; Thorpe, 1987b, 1994a; Lundqvist *et al.*, 1988; Hansen *et al.*, 1989b).

Maturation

First maturation may occur over a wide range of sizes (7 to 100 cm fork length), and of ages (0 to 8 years), and may be very variable within stocks (Saunders & Schom, 1985). It has been suggested that salmon must achieve certain threshold sizes before maturation is possible, but other than measures of minimum recorded length (e.g., Myers *et al.*, 1986), evidence for a limiting size is lacking. Indeed, from experimental tests Naevdal (1983) concluded that no minimal size need be attained before maturation can begin. Myers (1986) regarded the multiple age structure of spawning populations of Atlantic salmon as a mixed evolutionary stable strategy. For males he suggested that stable equilibria may exist between the proportion maturing as parr and proportions maturing after 1 or more years at sea that are related to the number of matings possible for the different categories of male. For females he suggested that such strategies were related to differences in the depths of spawning redd required by different sizes of females. Either case implies that rate of achievement of maturity is a heritable characteristic subject to selective forces.

Schaffer & Elson (1975), Power (1981) and Thorpe & Mitchell

(1981) showed that size at first spawning increased with length of upstream passage to spawning grounds, implying that large adult size (and late maturation) was an adaptive characteristic, selectively retained in populations of fish undergoing a difficult upstream migration. Scarnecchia (1983) found a similar relationship among stocks of 77 Icelandic rivers, but also noted relationships between June sea temperatures near the river mouths and age at spawning. He suggested that age at maturity was understood best in relation to each stock's entire life history pattern.

From cage rearing experiments Naevdal *et al.* (1978) and Saunders *et al.* (1983) showed that maturation age differed between stocks. Age at maturity is correlated negatively with growth rate (Alm, 1959), both phenotypically and genetically (Gjedrem, 1985). Researchers have shown that the proportion of a sibling population maturing at a given age under given growing conditions differs between families and stocks (Naevdal *et al.*, 1975, 1978; Thorpe, 1975, 1987a; Saunders & Sreedharan, 1977; Sutterlin & MacLean, 1984; Glebe & Saunders, 1986). Although several studies showed that the fastest-growing individuals mature first (Leyzerovich, 1973; Murphy, 1980; Saunders *et al.*, 1982; Thorpe *et al.,* 1983) and one failed to find such a relationship (Dempson *et al.*, 1986), the maturity process itself probably promotes growth during the early stages of maturation (Hunt *et al.*, 1982; Rowe & Thorpe, 1990a). Naevdal *et al.* (1983) found that maturing fish were heavier than immature fish of a given length, but did not find length growth differences between the two categories.

Individual fish make their developmental conversions in response to specific proximate cues. Photoperiod changes act as time-signals for such events, and Thorpe (1986) suggested that the physiological decision whether or not to mature is mediated internally using some biochemical index associated with growth rate or rate of acquisition of surplus energy (Rowe *et al.*, 1991). Genetic differences in the expression of maturity appear to depend on a threshold capability, awaiting the appropriate physiological-biochemical conditions, rather than a preset array of biochemical reactions and developments set to occur at a given age (Saunders, 1986; Thorpe, 1986). Apparent differences in life history strategy between river stocks may therefore be environmentally determined, and their genetic differences can only be assessed by comparative tests under identical rearing conditions.

Thorpe *et al.* (1983), Gjerde (1984a), Sutterlin & MacLean (1984) and Ritter *et al.* (1986) have all shown that parental age at maturity strongly influences maturation age of the progeny. Some investigators have claimed that maturity at the parr stage (in fresh water before smolt emigration) is a trait heritable independently from that of 'sea' age at maturity (Gjerde, 1984a; Gjedrem, 1985; Glebe & Saunders, 1986). Part of the problem in assessing such experiments lies in the lack of knowledge of whether the older adults themselves matured first as parr before entering the sea (Thorpe & Morgan, 1980). However, experimental evidence suggests that the difference is one of degree rather than kind (Adams & Thorpe, 1989). The age at maturity will be related to opportunity at the critical season (probably in autumn each year, when annually cyclic gonad growth begins [Thorpe, 1994b]), and there is no evidence that the processes of maturation at the parr stage and at the sea-run stage differ. The progeny of both types of adult are viable. However, investigators have also shown that the progeny of rapidly maturing fish (including mature parr) grow faster and mature earlier than the progeny of later maturing fish (Thorpe & Morgan, 1980; Glebe & Saunders, 1986).

Over a long series of experiments in Ireland, Piggins' (1983) data showed that the predominant influence on age at maturity was environmental. Of the progeny of parents that spawned after one or after two winters at sea, 98% and 87%, respectively, returned to spawn after one winter at sea. This consistent pattern suggested an underlying genetic control, heavily masked by environmental opportunity. Sutterlin *et al.* (1981) and Saunders *et al.* (1983) noted dramatic differences in proportions of fish maturing after 1 year at sea when part of a stock was caged and part was free ranging. They suggested that cold winter temperatures and/or reduced food intake in the cage localities inhibited maturation, which is consistent with the predictions of Thorpe's model (Thorpe, 1986). Subsequent experiments have shown that maturation can be arrested by food restriction over winter and spring, and that only those individuals that can replenish their fat stores in April will mature that year (Rowe & Thorpe, 1990b; Thorpe *et al.*, 1991; Reimers *et al.*, 1993). Jordan *et al.* (1990) showed a positive association between heterozygosity at the *MEP-2** locus and early maturity of salmon at sea. This finding was supported by similar evidence from a study in Spain (Moran *et al.*, 1994).

Return Timing and Rate

Nordqvist (1924) reviewed the timing of runs of adult salmon into rivers over their entire geographic range, and concluded that their seasonality was determined by local hydrological factors and was adjusted evolutionarily to ensure the fish were in the right place at the right time to overcome obstacles and reach the spawning grounds. Such a conclusion implied genetic control over developmental timing and migratory behaviour, which should be stock specific. Hansen & Jonsson (1991) provided experimental evidence for just such genetic control of seasonal return timing in different stocks of Norwegian salmon.

Within stocks, Ryman (1970), using data from smolt release experiments, found differences in recapture rates between half-sib families and reduced rates among inbred families. In a Canadian experiment in which smolts of three stocks were released in a series of places at differing distances from their native rivers, Ritter (1975) found that return rates showed a clinal decrease with distance of transplantation.

He interpreted this as a clinal decrease in sea survival, supporting a hypothesis that migration routes were heritable. In the pink salmon (*Oncorhynchus gorbuscha*), Bams (1976) showed that hybridization of introduced stock with local males restored a depressed return rate to a level close to that of the native fish. Bailey & Saunders (1984) found differences in return rate between stocks of Atlantic salmon released together in Canada. Ryman (1970) also noted that part of the variation in recapture frequencies between half-sib families was of additive genetic origin, and both he and Gjedrem (1986) have suggested that recapture frequency might be improved by selection. More recently, Bailey (1987) found indications of reductions in returns in successive generations of probably inbred stocks.

GAMETE PRODUCTION

Semen quantity in farmed salmon was significantly positively correlated with body weight and length (Gjerde, 1984b), but no evidence was presented on variation in this trait between stocks. Kazakov (1978a,b) found a similar increase in sperm production with fish size in the Neva stock, but this tended to decrease with repeated spawnings.

Sutterlin & MacLean (1984) recorded differences between stocks in investment in ovaries at ages 0 and 1, implying genetically controlled differences of potential fecundity. Chadwick *et al.* (1986) also found stock-specific differences in ovarian investment at the smolting stage, and these were correlated inversely with parental maturation age.

Pope *et al.* (1961) showed that fecundity increases with length of fish in six Scottish rivers, and that this relationship differed significantly between those six populations over 3 years. Aulstad & Gjedrem (1973) and Glebe *et al.* (1979) found stock differences in egg diameter in Norway and Canada, respectively, and Sutterlin & MacLean (1984) recorded differences between stocks in Newfoundland in numbers, sizes, and stage of development of previtellogenic oocytes at age 1 or more. Salmon egg size generally increases with parental size and egg number with growth rate (Thorpe *et al.*, 1984). Consequently, variation of these reproductive indices is closely related to growth performance. Gjedrem *et al.* (1986) found that egg volume and egg number were highly correlated genetically with body weight and length in 6 year-classes of salmon in Norway.

DISEASE RESISTANCE AND MALFORMATIONS

If disease resistance is genetically controlled, then reduced genetic variation may lead to increased susceptibility to infection (Cross & King, 1983). Gjedrem & Aulstad (1974) found significant differences in resistance to vibrio disease between river stocks of salmon in Norway, and Bailey (1986) demonstrated the heritability of resistance to furunculosis in Canadian salmon. Refstie (1986) and Fevolden *et al.* (1991) have also found heritable differences in stress responses, measured as blood cortisol and glucose levels.

Johnson & Jensen (1986, 1988, 1991) and Bakke (1991) have attributed recent substantial losses of salmon in some Norwegian rivers to infection by the monogenean trematode ectoparasite *Gyrodactylus salaris*. This parasite appears to have been introduced to Norwegian rivers by humans, on host salmon from the Baltic drainage, where the species is resistant to this parasite. The native Norwegian salmon populations, where the rivers drain into the Atlantic, show no such resistance. Hence the losses of these Norwegian salmon represent an indirect genetic effect of introducing parasite-tolerant carrier stocks of salmon from outside Norway into parasite-intolerant Norwegian salmon populations.

Susceptibility to a particular spinal deformity was found to be heritable, independently from heritable differences in growth and maturity (McKay & Gjerde, 1986).

ACIDITY

Decreased pH through acid precipitation has eliminated salmonids from a number of river systems throughout their range (Jensen & Snekvik, 1972; Leivestad *et al.*, 1976; Schom & Davidson, 1982). The changes from acceptable to lethal pH levels have been very rapid in relation to the generation time of the fish. Stock-specific and family differences in survival times of Atlantic salmon parr to low pH have been found experimentally (Schom & Saulnier, 1983; Schom, 1986), and suggest limited adaptation to stable local differences before recent pollution.

GENETIC CHANGE

The genetic constitution of populations is dynamic, and changes under processes that are directional (selection and gene flow) and random (drift). Taylor (1991) concluded that local adaptation was responsible for much of the genetic variation evident among salmonid populations. Thorpe & Koonce (1981) suggested that the Atlantic salmon, because of its divided stock structure, was particularly vulnerable to directional changes in genetic composition, and also to loss of diversity through loss of some of those stocks.

Human activities exert their greatest genetic effect on fish populations by selection through size-selective fisheries, habitat damage, and pollution (Wohlfarth, 1986). As these activities may affect the normal movements of salmon through overexploitation, construction of weirs or dams, alteration of habitat, or introduction of hatchery or exotic stocks, they also affect the pattern of gene flow between stocks.

Loss of diversity through drift may result from inbreeding, and is inversely proportional to twice the effective size of the breeding population (Falconer, 1981). Gene flow would normally counteract such a loss, and as some stocks of Atlantic salmon may number less than 40 breeding pairs, this source of genetic change is clearly very important.

However, Saunders & Schom (1985) have pointed out that in some stocks individual salmon from a single year-class may contribute to spawnings for as many as 9 years, and breed with salmon of 9 other year-classes. The effective breeding population would then be potentially much larger than the number of individuals present on the spawning grounds in any one season. Although this model has yet to be fully evaluated, Saunders & Schom suggest that this adaptive variation in life history strategy permits the persistence of such small stocks.

Major fisheries for salmon occur at sea, where the target populations are stock mixtures (Allendorf *et al.*, 1987). Piggins (1980) recorded that such fisheries reduced small stocks more severely than large ones, so that they may accelerate drift effects on the diversity of small stocks. Usually all female and most male spawners are of a size exploitable by the fisheries, so a stock's overall fitness could be severely reduced through the truncation of its spawning age structure by selective fishing of older, larger individuals.

Both size and age reductions have been documented in the Scottish and Canadian commercial catches (Schaffer & Elson, 1975; Ritter & Newbould, 1977; Porter *et al.*, 1986), and attributed to effects of the west Greenland fishery (Paloheimo & Elson, 1974). As male fish dominate in the younger age groups, truncation of the age range by the fishery influences the female component more severely. Schaffer (1974) and Wohlfarth (1986) have suggested that increased mortality of the older age classes should select for earlier age and smaller size at maturity. Caswell *et al.,* (1984) proposed that the recent increase in the proportion of maturing male parr in the Matamek River, Canada, was a consequence of just such a selection process. Alternatively, Myers *et al.,* (1986) attributed this change to an increased growth rate at lowered densities in freshwater.

In an important paper Riddell (1986) reviewed the problem critically, and suggested that the realized response to selection would be less than predicted from the unrealistically simple model of a single-trait response. He argued that such models did not account for genetic covariances, which were likely to be negative and would constrain responses to selection on correlated traits. Gjedrem (1985) and Thorpe (1986) have reported such negative genetic covariance between age at maturity and growth rate. Also, the partially tetraploid status of salmon calls into question predictions of their rate of response to selection of polygenic traits. Further, the harvested part of a population is inadequate for estimating selection intensity, whose net effects should be considered throughout the life history. Life history models developed to predict age at maturity are more sensitive to juvenile than to adult mortality (Jonsson *et al.*, 1984; Stearns & Crandall, 1984), so heavy fishery exploitation is more likely to reduce density-dependent factors on juveniles than to affect age at maturity genetically. Overall, Riddell concluded that while excessive fishing can reduce spawning abundances and may increase the incidence of parr maturity, these responses may not involve genetic change in maturation age. However, in a similar situation in Kamchatka, genetic change does appear to have occurred in a heavily exploited sockeye salmon stock (*Oncorhynchus nerka*) (Thorpe, 1994c), in which maturation, occurring at progressively younger ages over a 50-year period, has been accompanied by an increase in heterozygosity.

Habitat alteration, especially through the total exclusion of salmon from spawning and nursery grounds by dams (Gray, 1974; Ros, 1981; Vuorinen, 1982; Jones, 1988), and more recently by acidification (Saunders, 1981; Schom & Davidson, 1982), can be more drastic in its effects than fisheries, and has accounted for the total loss of many salmon stocks (Saunders, 1981). Attempts to restore such populations show that important genetic adaptations were lost with those stocks (Jones, 1988). Physiological adaptation and developmental timing differ between stocks, particularly in relation to migrations (Riddell & Leggett, 1981; Riddell *et al.*, 1981; Saunders, 1981). Restoration attempts have revealed the close relationship between rearing locality and developmental timing (e.g., Struthers & Stewart, 1986), and emphasize the likely loss of unique genetic resources following habitat alteration.

While habitat change through industrial, domestic, and agricultural pollution produces obvious and often catastrophic effects on salmon stocks, more sinister genetic changes may be induced through deliberate or accidental mixtures of stocks. Enhancement of wild populations by introduction of hatchery stocks has been widespread (MacCrimmon & Gots, 1979). Currently in Norway alone, 10 million to 15 million fry are hatched annually to enhance wild stocks (Hansen *et al.*, 1987a). In addition, an unknown quantity is escaping from the rapidly growing farming industry (Lund *et al.*, 1991). As a conservative estimate this may amount to 1% of annual production, equivalent to 500 tonnes of escapees in 1987 (Hansen *et al.*, 1987a). The escapees from sea cages are known to be 'homeless', and to enter a wide range of rivers at maturity (Hansen *et al.*, 1987b). Gausen (1988) found an average of 13% escapees among wild salmon collected for broodstock in 1987, and a survey of 16 Norwegian rivers during the spawning season of 1989 yielded 698 reared fish out of 1,791 examined (39%) (Lund *et al.*, 1991). Lura & Sægrov (1991) recorded successful spawning by farm escapees in the wild in Norway, as did Webb *et al.* (1991) in Scotland. Farm escapees, including spawned fish, have also been recorded in rivers in Iceland (Gudjonsson, 1991).

Selection for particular traits is usually gained at the expense of genetic variance and results in decreased fitness (e.g., Altukhov & Salmenkova, 1991). Cross & King (1983) and Cross & Challanain (1991) found that gene frequencies differed significantly at a number of loci between wild and hatchery samples. Ståhl (1987), Verspoor (1988) and Kazakov & Titov (1993) showed that some hatchery stocks of Atlantic salmon were genetically much less variable than wild stocks, and they attributed this to the use of too few parental fish. Ovenden *et al.*, (1993) found no restriction site variation at all among mitochondrial DNA samples of Tasmanian Atlantic salmon, and attributed this to the transitory decrease in numbers of broodfish in the hatchery from which the Tasmanian population was derived. However, comparing allozyme and mitochondrial DNA variation in strains of farmed salmon in Norway, Hindar & Jonasson (1990) found variation within and between strains comparable to that among wild stocks. Hindar (K. Hindar, Directorate of Nature Management, Trondheim, Norway, pers. comm., 1988) believes that this high genetic diversity is due to the original systematic development of these farmed strains at the breeding station at Sunndalsøra from several wild stocks, and to the routine use of high numbers of parent fish (120 families in each case) (Gjedrem *et al.*, 1988, 1991). In the most popular farmed line in Ireland, mean heterozygosity and mean number of alleles per locus were comparable to values in wild salmon populations (Cross & Challanain, 1991). Youngson *et al.* (1991) also found no difference in mean heterozygosity between farmed and wild stocks sampled in Scotland in 1988 and 1989. Ståhl (1987) showed that stock differentiation occurred within as well as between river systems, and warned against the breakdown of adapted gene complexes (and reduction of diversity) that might result from hybridization with nonadapted, introduced stocks. While evidence for such introgression in Atlantic salmon populations is scarce at present, the risk of introgression occurring increases with the growth of salmon farming (Saunders, 1991), which until recently has been increasing at about 50% per year. Hindar *et al.*, (1991) have reviewed this topic comprehensively. Hutchings (1991) modelled the influence of spawning intrusions of farmed fish on population size in wild Atlantic salmon, by varying the proportional representation of cultured salmon at spawning, the frequency of spawning intrusions, and the difference in fitness between the two forms for three mating situations: no interbreeding, successful interbreeding, and interbreeding with hybrid dysgenesis. The extinction probability of native genomes was greatest when interbreeding occurred in relatively large numbers, was greatest for small populations, and was also influenced by nongenetic (increased density) factors. Mork (1991) drew attention to the particular genetic risks associated with 'burst' immigrations of farmed fish into wild populations, following the wrecks of cage farms in storms [cf. Webb *et al.*, (1991) who reported an accident in which 184,000 maturing salmon escaped from a cage farm in a storm, in a coastal inlet in the neighbourhood of a small river, and some subsequently spawned in that river]. Using known allele frequencies at three loci and population sizes of five wild Norwegian stocks and one farmed stock, Mork calculated potential gene flow and estimated potential reduction in genetic differentiation between the stocks as a result of such catastrophic immigrations. Because of the large numbers of immigrants, the greatest genetic effect of such events is expressed in the first generation. Assuming equal fitness of immigrants and natives, and a 30% representation of immigrants at spawning, he estimated reductions in genetic differentiation of between 50-70% in one generation.

Lack of detailed knowledge of the genetic structure of particular wild stocks vulnerable to interference from cultured fishes hampers the assessment of levels of introgression, but Knox & Verspoor (1991) have described a mitochondrial DNA restriction fragment length polymorphism unique to Norwegian salmon which may prove a useful marker in the future. Even more precise fingerprinting techniques are being developed, with the recent synthesis of a novel DNA probe (Ssal-rep) that gives clear highly polymorphic fingerprints for salmonids (Bentzen *et al.*, 1993).

GENETIC CONSERVATION

Most experts agree about the strong need to conserve existing stocks, as these represent a source of genetic diversity for future use (e.g., Ryman, 1981). Methods for achieving this have been both *in situ* and *ex situ.*

In Situ Conservation

Catch quotas and restrictions on fishing gear, times, and seasons are all part of the stock-in-trade of national management programmes for the salmon resource. While these all go some way toward protecting the gene pools of stocks, debate persists on the extent to which even restricted fisheries reduce genetic diversity. Canada recently closed some commercial fisheries because of the apparently endangered state of many stocks.

A Swedish report (Nyman & Norman, 1987) provides a basis for a national strategy for conserving salmon genetic resources in that country, where most rivers have been modified extensively for hydroelectric power production, and where stocking of hatchery fish is massive and necessary to compensate for the loss of natural production. For such modified rivers, the report recommends using river-specific stocks, from which the progeny of at least 25 pairs to minimize reduction of genetic diversity, chosen at random from the whole migration period, should be mixed and released as tagged smolts over several years. In extreme cases where less than 25 pairs of broodfish are available, the report recommends using more than 25 of the commoner sex. In such cases, the progeny from each pairing should be culled to equal numbers. (Waples *et al.* [1990] suggested a more cautious approach, noting the dependence of effective number of salmonid parents on generation time, and suggested that broodfish numbers should be in the hundreds of pairs.) The report emphasizes that cultured fish should not be released into rivers that are not modified for power production, and that those rivers that can support wild populations should be used as gene banks.

A Canadian report (Anderson, 1987) follows similar lines, but is not so explicitly genetically oriented. However, its recommendations for criteria for broodstock choice in enhancement programmes in Newfoundland take full account of the stock structure of salmon populations and their adaptive nature.

Ståhl (1987) has questioned the wisdom of basing conservation policies on whole rivers as the stock unit, as he and others have shown that at least some rivers have multiple separate breeding units. Treating them all as one ultimately would break down the isolating mechanisms between them and lead to loss of diversity.

Ex Situ Conservation

Cryopreservation has been developed successfully for storing sperm (Scott & Baynes, 1980; Stoss & Refstie, 1982), but not for eggs (McAndrew *et al.*, 1993). Methods for storing eggs or zygotes are still under investigation (B.McAndrew, University of Stirling, Scotland, pers. comm.).

In 1986 the Norwegian Directorate for Nature Management began a programme of cryopreservation of Atlantic salmon sperm and by 1990 had collected sperm from 2070 individuals from 120 separate stocks (Bergan *et al.*, 1991). Their goal is to obtain sperm from at least 50 males from each stock, and to increase the number of stocks represented to 150. In 1990 they established a living gene bank (hatchery-maintained populations) of 74 families from 10 stocks. It is intended to increase the number of stocks and families, to ensure a capacity for producing eggs, for security, re-establishment, or enhancement of 20-30 threatened stocks in Central Norway.

CONCLUSION

In recent studies prompted by the need to register threatened stocks of Pacific salmon under the Endangered Species Act in the USA, the following statement was made (Riggs, 1990: quoted in Waples, 1991):

> Given that (a) Pacific salmon represent a unique resource of great cultural, social, and economic value, and that (b) this resource is the product of many thousands of years of evolution, therefore responsible public policy should not be driven solely by short-term considerations. Rather, as stewards of the resource, all responsible parties should seek ways to ensure that the resource is available for many generations to come. Purposeful planning to assure the quality and effectiveness of production activities must acknowledge that sustained production and long-term health of the resource can be achieved only by conserving the genetic information in this evolutionary legacy.

This statement is equally applicable and appropriate to the need for conservation of the genetic resource represented by Atlantic salmon.

REFERENCES

Adams, C.E. & Thorpe, J.E. 1989. Photoperiod and temperature effects on early development and reproductive investment in Atlantic salmon (*Salmo salar* L.). *Aquaculture* **79**, 403-409.

Allendorf, F.W., Ryman, N. & Utter, F.M. 1987. Genetics and fisheries management, past, present and future. In *Population Genetics and Fishery Management* (Ryman, N. & Utter, F. eds.) pp.1-19. Seattle: University of Washington Press.

Alm, G. 1959. Connection between maturity, size and age in fishes. *Report of the nstitute of Freshwater Research (Drottningholm)* **40**, 5-145.

Altukhov, Yu.P. & Salmenkova, E.A. 1987. Population genetics of cold water fish. In *Selection, Hybridisation and Genetic Engineering in Aquaculture* (Tiews, K. ed.) Vol.I pp.3-29. Berlin: Heenemann.

Altukhov, Yu.P. & Salmenkova, E.A. 1991. The genetic structure of salmon populations. *Aquaculture* **98**, 11-40.

Anderson, T.C. 1987. Broodstock acquisition for Atlantic salmon (*Salmo salar)* enhancement activities in Newfoundland: A review of related factors and suggested criteria. *Canadian Manuscript Reports in Fisheries & Aquatic Sciences* **1931**.

Antere, I. & Ikonen, E. 1983. A method of distinguishing wild salmon from those originating from fish farms on the basis of scale structure. *International Council for the Exploration of the Sea* **CM.1983/M:26**.

Aulstad, D. & Gjedrem, T. 1973. The egg size of salmon (*Salmo salar*) in Norwegian rivers. *Aquaculture* **2**, 337-341.

Baglinière, J.-L. & Champigneulle, A. 1986. Population estimates of juvenile Atlantic salmon, *Salmo salar*, as indices of smolt production in the R. Scorff, Brittany. *Journal of Fish Biology* **29**, 467-482.

Baglinière, J-L. & Maisse, G. 1985. Precocious maturation and smoltification in wild Atlantic salmon in the Armorican Massif, France. *Aquaculture* **45**, 249-263.

Bailey, J.K. 1986. Differential survival among full sib Atlantic salmon (*Salmo salar*) families exposed to furunculosis. *Salmon Genetics Research Program Technical Report Series* **58**. New Brunswick, Canada: North American Salmon Research Center.

Bailey, J.K. 1987. Canadian sea ranching program (East Coast). In *Selection, Hybridisation and Genetic Engineering in Aquaculture* (Tiews, K. ed.) Vol. II, pp.443-448. Berlin: Heenemann.

Bailey, J.K. & Saunders, R.L. 1984. Returns of three year-classes of sea-ranched Atlantic salmon of various river strains and strain crosses. *Aquaculture* **41**, 259-270.

Bailey, J.K., Saunders, R.L. & Buzeta, M.I. 1980. Influence of parental smolt age and sea age on growth and smolting of hatchery-reared Atlantic salmon (*Salmo salar*). *Canadian Journal of Fisheries & Aquatic Sciences* **37**, 1379-1386.

Bakke, T.A. 1991. A review of inter- and intraspecific variability in salmonid hosts to laboratory infections with *Gyrodactylus salaris* Malmberg. *Aquaculture* **98**, 303-310

Bams, R.A. 1976. Survival and propensity for homing as affected by presence or absence of locally adapted paternal genes in two transplanted populations of pink salmon (*Oncorhynchus gorbuscha*). *Journal of the Fisheries Research Board of Canada* **33**, 2716-2725.

Barsiene, J.V. 1981. Intercellular polymorphism of chromosome sets in the Atlantic salmon. *Tsitologiya* **23**, 1053-1059.

Behnke, R.J. 1972. The systematics of salmonid fishes of recently glaciated lakes. *Journal of the Fisheries Research Board of Canada* **29**, 639-671.

Bentzen, P., Taylor, E.B. & Wright, J.M. 1993. A novel synthetic probe for DNA fingerprinting salmonid fishes. *Journal of Fish Biology* **43**, 313-316.

Bergan, P.I., Gausen, D. & Hansen, L.-P. 1991. Attempts to reduce the impact of reared Atlantic salmon on wild in Norway. *Aquaculture* **98,** 319-324.

Bermingham, E., Forbes, S.H., Friedland, K. & Pla, C. 1991. Discrimination between Atlantic salmon (*Salmo salar*) of North American and European origin using restriction analyses of mitochondrial DNA. *Canadian Journal of Fisheries & Aquatic Sciences* **48**, 884-893.

Bern, H.A. & Mahnken, C.V.W. eds. 1982. Salmon smoltification. *Aquaculture* **28**, 1-270.

Beverley-Burton, M. 1978. Population genetics of (*Anisakis simplex*) (Nematoda: Ascaridoidea) in Atlantic salmon (*Salmo salar*) and their use as biological indicators of host stocks. *Environmental Biology of Fishes* **3**, 369-377.

Beverley-Burton, M. & Pippy, J.H.C. 1978. Distribution, prevalence and mean numbers of larval *Anisakis simplex* (Nematoda: Ascaridoidea) in Atlantic salmon (*Salmo salar*) and their use as biological indicators of host stocks. *Environmental Biology of Fishes* **3**, 211-222.

Birt, T.P., Green, J.M. & Davidson, W.S. 1986. Analysis of mitochondrial DNA in allopatric anadromous and nonanadromous Atlantic salmon, *Salmo salar*. *Canadian Journal of Zoology* **64**, 118-120.

Birt, T.P., Green, J.M. & Davidson, W.S. 1991a. Contrasts in development and smolting of genetically distinct sympatric anadromous and non-anadromous Atlantic salmon, *Salmo salar*. *Canadian Journal of Zoology* **69**, 2075-2084.

Birt, T.P., Green, J.M. & Davidson, W.S. 1991b. Mitochondrial DNA variation reveals genetically distinct sympatric populations of anadromous and non-anadromous Atlantic salmon, *Salmo salar*. *Canadian Journal of Fisheries & Aquatic Sciences* **48**, 577-582

Boothroyd, E.R. 1959. Chromosome studies on three Canadian populations of Atlantic salmon, *Salmo salar* L. *Canadian Journal of Genetics & Cytology* **1**, 161-172.

Caswell, H., Naiman, R.J. & Morin, R. 1984. Evaluating the consequences of reproduction in complex salmonid life cycles. *Aquaculture* **43**, 123-134.

Chadwick, E.M.P., Randall, R.G. & Leger, C. 1986. Ovarian development of Atlantic salmon (*Salmo salar*) smolts and age at first maturity. In *Salmonid Age at Maturity* (Meerburg, D.J. ed.). *Canadian Special Publication in Fisheries & Aquatic Sciences* **89**, 15-23.

Couturier, C.Y., Clarke, L. & Sutterlin, A.M. 1986. Identification of spawning areas of two forms of Atlantic salmon *Salmo salar* inhabiting the same watershed. *Fisheries Research* **4**, 131-144.

Cross, T.F. & Challanain, D.N. 1991. Genetic characterisation of Atlantic salmon (*Salmo salar*) lines farmed in Ireland. *Aquaculture* **98**, 209-216.

Cross, T.F. & Healy, J.A. 1983. The use of biochemical genetics to distinguish populations of Atlantic salmon, *Salmo salar*. *Irish Fisheries Investigations Series A* **23**.

Cross, T.F. & King, J. 1983. Genetic effects of hatchery rearing in Atlantic salmon. *Aquaculture* **33**, 33-40.

Cross, T.F. & Payne, R.H. 1977. NADP-isocitrate dehydrogenase polymorphism in the Atlantic salmon *Salmo salar*. *Journal of Fish Biology* **11**, 493-496.

Cross, T.F. & Ward, R.D. 1980. Protein variation and duplicate loci in the Atlantic salmon (*Salmo salar*). *Genetic Research, Cambridge* **36**, 147-165.

Cross, T.F., Healy, J.A. & O'Rourke, F.J. 1978. Population discrimination in Atlantic salmon from Irish rivers using biochemical genetic methods. *International Council for the Exploration of the Sea* **CM.1987/M:2**.

Cutler, M.G., Bartlett, S.E., Hartley, S.E. & Davidson, W.S. 1991. A polymorphism in the ribosomal RNA genes distinguishes Atlantic salmon (*Salmo salar*) from North America and Europe. *Canadian Journal of Fisheries & Aquatic Sciences* **48**, 1655-1661.

Davidson, W.S., Green, J.M. & Birt, T.P. 1989a. A review of genetic variation in Atlantic salmon, *Salmo salar* L., and its importance for stock identification, enhancement programmes and aquaculture. *Journal of Fish Biology* **34**, 547-560.

Davidson, W.S., Birt, T.P. & Green, J.M. 1989b. Organization of the mitochondrial genome from Atlantic salmon (*Salmo salar*). *Genome* 32, 340-342.

Dempson, J.B., Myers, R.A. & Reddin, D.G. 1986. Age at first maturity of Atlantic salmon (*Salmo salar*): Influences of the marine environment. In *Salmonid Age at Maturity* (Meerburg, D.J. ed.) *Canadian Special Publications in Fisheries & Aquatic Sciences* **89**, 79-89

Falconer, D.S. 1981. *Introduction to Quantitative Genetics*. 2d. ed. London: Longman.

Fevolden, S.E., Refstie, T. & Roed, K.H. 1991. Selection for high and low cortisol stress in Atlantic salmon (*Salmo salar*) and rainbow trout (*Oncorhynchus mykiss*). *Aquaculture* **95**, 53-65.

Friars, G.W., Bailey, J.K. & Saunders, R.L. 1979. Consideration of a method of analyzing diallel crosses of Atlantic salmon. *Canadian Journal of Genetics & Cytology* **21**,121-128.

Gausen, G. 1988. Registreringer av oppdrettslaks i vassdrag. In *Fagmote om sikringssoner for laksefisk* (Lindgren, B.ed.) pp.58-69. Norway: Stjordal.

Gjedrem, T. 1985. Genetic variation in age at maturity and its relation to growth rate. In*Salmonid Reproduction* (Iwamoto, R.N. & Sower, S. eds.) pp.52-61. Seattle: University of Washington Press.

Gjedrem, T. 1986. Breeding plan for sea ranching. *Aquaculture* **57**,77-80

Gjedrem, T. & Aulstad, D. 1974. Selection experiments with salmon. I: Differences in resistance to vibrio disease of salmon parr (*Salmo salar*). *Aquaculture* **3**, 51-59.

Gjedrem, T., Eggum, A. & Refstie, T. 1977. Chromosomes of some salmonids and salmonid hybrids. *Aquaculture* **11**, 335-348.

Gjedrem, T., Gjerde, B. & Refstie, T. 1988. A review of quantitative genetic research in salmonids at AKVAFORSK. In *Proceedings of the second International Conference on Quantitative Genetics* (Weir, B.S., Eisen, E.J., Goodman, M.M. & Namkoong, G. eds.) pp. 527-535. Sunderland MA: Sinauer.

Gjedrem, T., Gjoen, H.M. & Gjerde, B. 1991. Genetic origin of Norwegian farmed salmon. *Aquaculture* **9**, 41-50.

Gjedrem, T., Haus, E. & Halseth, V. 1986. Genetic variation in reproductive traits in Atlantic salmon and rainbow trout. *Aquaculture* **57**, 369.

Gjerde, B. 1984a. Response to individual selection for age at sexual maturity in Atlantic salmon. *Aquaculture* **33**, 51-72.

Gjerde, B. 1984b. Variation in semen production of farmed Atlantic salmon and rainbow trout. *Aquaculture* **40**, 109-114.

Gjerde, B. & Gjedrem, T. 1984. Estimates of phenotypic and genetic parameters for carcass traits in Atlantic salmon and rainbow trout. *Aquaculture* **36**, 97-110.

Glebe, B.D. & Saunders, R.L. 1986. Genetic factors in sexual maturity of cultured Atlantic salmon (*Salmo salar*) parr and adults reared in sea cages. In *Salmonid Age at Maturity* (Meerburg, D.J. ed.) *Canadian Special Publications in Fisheries & Aquatic Sciences* **89**, 24-29

Glebe, B.D., Appy, T.D. & Saunders, R.L. 1979. Variation in Atlantic salmon (*Salmo alar*) reproductive traits and their implications in breeding programs. *International Council for the Exploration of the Sea* **CM.1979/M:23**.

Grammeltvedt, A.F. 1975. Chromosomes of salmon (*Salmo salar*) by leucocyte culture. *Aquaculture* 5, 205-209.

Gray, R.W. 1974. Salmon development on the LaHave River. *Atlantic Salmon Journal* **1**,14-17.

Gudjonsson, S. 1991. Occurrence of reared salmon in natural salmon rivers in Iceland. *Aquaculture* **98**, 133-142.

Gunnes, K. 1979. Survival and development of Atlantic salmon eggs and fry at three different temperatures. *Aquaculture* **16**, 211-218.

Gunnes, K. & Gjedrem, T. 1978. Selection experiments with salmon IV. Growth of Atlantic salmon during two years in the sea. *Aquaculture* **15**, 19-33.

Gyllensten, U. 1985. The genetic structure of fish: differences in the intraspecific distribution of biochemical genetic variation between marine, anadromous, and freshwater species. *Journal of Fish Biology* **26**, 691-699.

Gyllensten, U. & Wilson, A.C. 1987. Mitochondrial DNA of Salmonids. In *Population Genetics and Fishery Management* (Ryman, N. & Utter, F. eds.) pp.301-318. Seattle: University of Washington Press.

Hansen, L.-P. & Jonsson, B. 1991. Evidence of a genetic component in the seasonal return pattern of the Atlantic salmon, *Salmo salar*. L. *Journal of Fish Biology* **38**, 251-258

Hansen, L.-P. & Lea, T. 1982. Tagging and release of Atlantic salmon smolts (*Salmo alar* L.) in the River Rana, Northern Norway. *Report of the Institute of Freshwater Research (Drottningholm)* **60**, 31-38.

Hansen, L.-P., Doving, K.B. & Jonsson, B. 1987a. Migration of farmed adult Atlantic salmon with and without olfactory sense, released on the Norwegian coast. *Journal of Fish Biology* **30**, 713-721.

Hansen, L.-P., Lund, R.A. & Hindar, K. 1987b. Possible interaction between wild and reared Atlantic salmon in Norway. *International Council for the Exploration of the Sea* **CM.1987/M:14**.

Hansen, L.-P., Clarke, W.C., Saunders, R.L. & Thorpe, J.E. eds. 1989a. Salmon Smolting Workshop III. *Aquaculture* **82**,1-390.

Hansen, L.-P., Jonsson, B., Morgan, R.I.G. & Thorpe, J.E. 1989b. Influence of parr maturity on emigration of smolting Atlantic salmon (*Salmo salar*). *Canadian Journal of Fisheries & Aquatic Sciences* **46**, 410-415.

Hartley, S.E. 1987. The chromosomes of salmonid fishes. *Biological Reviews* **62**, 197-214.

Hawkins, A.D., Urquhart, G.G. & Shearer, W.M. 1979. The coastal movements of returning Atlantic salmon, *Salmo salar* (L.). *Scottish Fisheries Research Report* **15**.

Heggberget, T., Lund, R.A., Ryman, N. & Ståhl, G. 1986. Growth and genetic variation of Atlantic salmon (*Salmo salar*) from different sections of the River Alta, North Norway. *Canadian Journal of Fisheries & Aquatic Sciences* **43**, 1828-1835.

Higgins, P.J. 1985. Metabolic differences between Atlantic salmon (*Salmo salar)* parr and smolts. *Aquaculture* **45**, 33-53.

Higgins, P.J. & Talbot, C. 1985. Growth and feeding in juvenile Atlantic salmon (*Salmo salar L.*). In *Nutrition and Feeding in Fish* (Cowey, C.B., Mackie, A.M. & Bell, J.G. eds.) pp.243-263. London: Academic Press.

Hindar, K. & Jonasson, J.E. 1990. A comparison of allozyme and mitochondrial

DNA variation in Atlantic salmon (*Salmo salar*). *Aquaculture* **85**, 330-331.

Hindar, K., Ryman, N. & Utter, F. 1991. Genetic effects of cultured fish on natural fish populations. *Canadian Journal of Fisheries & Aquatic Sciences* **48,** 945-957.

Hovey, S.J., King, D.P.F., Thompson, D. & Scott, A. 1989. Mitochondrial DNA and allozyme analysis of Atlantic salmon, *Salmo salar* L., in England and Wales. *Journal of Fish Biology* **35 (Supplement A)**, 253-260.

Hunt, S.M.V., Simpson, T.H. & Wright, R.S. 1982. Seasonal changes in the levels of 11-oxotestosterone and testosterone in the serum of male salmon, *Salmo salar L.*, and their relationship to growth and maturation cycle. *Journal of Fish Biology* **20**, 105-119.

Huntingford, F.A., Thorpe, J.E., Garcia de Leaniz, C. & Hay, D.W. 1992. Patterns of growth and smolting in autumn migrants from a Scottish population of Atlantic salmon, *Salmo salar* L. *Journal of Fish Biology* **41 (Supplement B)**, 43-51.

Hurley, S.M. & Schom, C.B. 1984. Genetic control of swimming stamina in Atlantic salmon *Salmo salar*. *Canadian Journal of Genetics & Cytology* **26**, 57-61.

Hutchings, J.A. 1991. The threat of extinction to native populations experiencing spawning intrusions by cultured Atlantic salmon. *Aquaculture* **98**, 119-132.

Ihssen, P.E., Booke, H.E., Casselman, J.M., McGlade, J.M., Payne, N.R. & Utter, F.M. 1981. Stock identification: materials and methods. *Canadian Journal of Fisheries & Aquatic Sciences* **38**, 1838-1855.

Jensen, A.J. & Johnsen, B.O. 1986. Different adaptation strategies of Atlantic salmon (*Salmo salar*) populations to extreme climates with special reference to some cold Norwegian rivers. *Canadian Journal of Fisheries & Aquatic Sciences* **43**, 980-984.

Jensen, J.M. & Lear, W.H. 1980. Atlantic salmon caught in the Irminger Sea and at east Greenland. *Journal of Northwest Atlantic Fisheries Science* **1**, 55-64.

Jensen, K.W. & Snekvik, E. 1972. Low pH levels wipe out salmon and trout populations in southern Norway. *Ambio* **1**, 223-225.

Johnsen, B.O. & Jensen, A.J. 1986. Infestations of Atlantic salmon, *Salmo salar* by*Gyrodactylus salaris* in Norwegian rivers. *Journal of Fish Biology* **29**, 233-276.

Johnsen, B.O. & Jensen, A.J. 1988. Introduction and establishment of *Gyrodactylus salaris* Malmberg, 1957, on Atlantic salmon, *Salmo salar L.* fry and parr in the River Vefsna, northern Norway. *Journal of Fish Diseases* **11,** 35-45.

Johnsen, B.O. & Jensen, A.J. 1991. The *Gyrodactylus* story in Norway. *Aquaculture* **98**, 289-302.

Jones, R.A. 1988. Atlantic salmon restoration in the Connecticut River. In *Atlantic Salmon: Planning for the Future* (Mills, D.H. & Piggins, D.J. eds.) pp.415-426 London: Croom Helm.

Jonsson, B., Hindar, K. & Northcote, T.G. 1984. Optimal age at sexual maturity of sympatric and experimentally allopatric cutthroat trout and Dolly Varden charr. *Oecologia* **61**, 319-325.

.Jordan, W.C. & Youngson, A.F. 1991. Genetic protein variation and natural selection in Atlantic salmon (*Salmo salar* L.) parr. *Journal of Fish Biology* **39 (Supplement A)**, 185-192.

Jordan, W.C., Youngson, A.F. & Webb, J.H. 1990. Genetic variation at the malic enzyme-2 locus and age at maturity in sea-run Atlantic salmon (*Salmo salar*). *Canadian Journal of Fisheries & Aquatic Sciences* **47**, 1672-1677.

Kazakov, R.V. 1978a. Izmenenie kachestva polovykh produktov samtsov atlanticheskogo lososya Nevskoyi populyatsii vo vremya neresta. *Izvestia GosNIORKh* **129**, 85-93.

Kazakov, R.V. 1978b. Nekotorye osobennosti produtsirovaniya spermy i zavisimost eye kachestva ot vozrasta, vesa i kratnosti uchastiya v nereste samtsov atlanticheskogo lososya. *Izvestia GosNIORKh* **129**, 94-102.

Kazakov, R.V. & Titov, S.F. 1991. Geographical patterns in the population genetics of Atlantic salmon, *Salmo salar* L., on U.S.S.R. territory, as evidence for colonization routes. *Journal of Fish Biology*. **39**, 1-6.

Kazakov, R.V. & Titov, S.F. 1993. Population genetcis of salmon *Salmo salar* L., in northern Russia. *Aquaculture & Fisheries Management* **24**, 495-506.

Khanna, N.D., Juneja, R.K., Larsson, B. & Gahne, B. 1975. Electrophoretic studies on proteins and enzymes in Atlantic salmon (*Salmo salar* L.). *Swedish Journal of Agricultural Research* **5**, 185-192.

Knox, D. & Verspoor, E. 1991. A mitochondrial DNA restriction fragment length polymorphism of potential use for discrimination of farmed Norwegian and wild Atlantic salmon populations in Scotland. *Aquaculture* **98**, 249-257.

Koljonen, M.-L. 1989. Electrophoretic genetic variation in natural and hatchery stocks of Atlantic salmon in Finland. *Hereditas* **110**, 23-35.

Kristinsson, J.B., Saunders, R.L. & Wiggs, A.J. 1985. Growth dynamics during the development of bimodal length-frequency distribution in juvenile Atlantic salmon (*Salmo salar* L.). *Aquaculture* **45**, 1-20.

Kwain, W. & Lawrie, A.H. 1981. Pink salmon in the Great Lakes. *Fisheries* **6(2)**, 2-6.

Larsson, P-O. 1974. Migration of the Swedish west coast salmon stocks. *Laxforskningsinstitutet Meddelande* **3.**

Lear, W.H. & Sandeman, E.J. 1980. Use of scale characters and discriminant functions for identifying continental origin of Atlantic salmon. *Rapports et proces-verbaux, Réunion Conseil Permanent International pour l'Exploration de la Mer* **176**, 68-75.

Leivestad, H., Hendrey, G., Muniz, I.P. & Snekvik, E. 1976. Effect of acid precipitation on freshwater organisms. In *Impact of Acid Precipitation on Forest and Freshwater Ecosystems in Norway* (Braekke, F.H. ed.) SNSF Project **FR6/76**, 87-111. Ås, Norway: Norwegian Council for Scientific Research.

Leyzerovich, K.A. 1973. Dwarf males in hatchery propagation of the Atlantic salmon. *Journal of Ichthyology* **13**, 382-391.

Lund, R. A., Økland, F. & Hansen, L.-P. 1991. Farmed Atlantic salmon (*Salmo salar*) in fisheries and rivers in Norway. *Aquaculture* **98**, 143-150.

Lundqvist, H., Clarke, W.C. & Johansson, H. 1988. The influence of precocious sexual maturation on survival to adulthood of river stocked Baltic salmon (*Salmo salar*) smolts. *Holarctic Ecology* **11**, 60-69.

Lura, H. & Sægrov, H. 1991. Documentation of sucessful spawning of escaped farmed female Atlantic salmon, *Salmo Salar*, in Norwegian rivers. *Aquaculture* **98**,151-159.

MacCrimmon, H.R. & Claytor, R.R. 1985. Meristic and morphometric identity of Baltic stocks of Atlantic salmon (*Salmo salar*). *Canadian Journal of Zoology* **63**, 2032-2037.

MacCrimmon, H.R. & Claytor, R.R. 1986. Possible use of taxonomic characters to identify Newfoundland and Scottish stocks of Atlantic salmon, *Salmo salar* L. *Aquaculture & Fisheries Management.* **17**, 1-17.

MacCrimmon, H.R. & Gots, B.L. 1979. World distribution of Atlantic salmon. *Journal of the Fisheries Research Board of Canada* **36**, 422-457.

McAndrew, B.J., Rana, K.J. & Penman, D.J. 1993. Conservation and preservation of genetic variation in aquatic organisms. In *Recent Advances in Aquaculture* (Muir, J.F. & Roberts, R.J. eds.) pp.295-336. Oxford: Blackwells.

McDowall, R.M. 1990. *New Zealand Freshwater Fishes.* Auckland: Heinemann Reed.

McElligott, E.A. & Cross, T.F. 1991. Protein variation in wild Atlantic salmon, with particular reference to southern Ireland. *Journal of Fish Biology* **39 (Supplement A)**, 35-42.

McKay, L.R. & Gjerde, B. 1986. Genetic variation for a spinal deformity in Atlantic salmon *Salmo salar*. *Aquaculture* **52**, 263-272.

McKenzie, J.A. & Paim, U. 1969. Variations in the plasma proteins of Atlantic salmon *Salmo salar*. *Canadian Journal of Zoology* **47**, 759-761.

McVeigh, H.P., Bartlett, S.E. & Davidson, W.S. 1991. Polymerase chain reaction/direct sequence analysis of the cytochrome b gene in *Salmo salar*. *Aquaculture* **95**, 225-233.

Metcalfe, N.B. & Thorpe, J.E. 1990. Determinants of geographical variation in the age of seaward migrating salmon, *Salmo salar*. *Journal of Animal Ecology* **59**, 135-145.

Metcalfe, N.B., Huntingford, F.A. & Thorpe, J.E. 1986. Seasonal changes in feeding motivation of juvenile Atlantic salmon (*Salmo salar*). *Canadian Journal of Zoology* **64**, 2439-2446.

Metcalfe, N.B., Huntingford, F.A., Graham, W.D. & Thorpe, J.E. 1989. Early social status and the development of life-history strategies in Atlantic salmon. *Proceedings of the Royal Society of London* **B236**, 7-19.

Milner, A.M. 1987. Colonization and ecological development of new streams in Glacier Bay National Park, Alaska. *Freshwater Biology* **18**, 53-70.

Milner, A.M. & Bailey, R.G. 1989. Salmonid colonization of new streams in Glacier Bay National Park, Alaska. *Aquaculture & Fisheries Management* **20**, 179-192.

Møller, D. 1970. Transferrin polymorphism in Atlantic salmon (*Salmo salar*). *Journal of the Fisheries Research Board of Canada* **27**, 1617-1625.

Moran, P., Pendas, A.M., Garcia-Vazquez, E., Izquierdo, J.T. & Rutherford, D.T. 1994. Electrophoretic assessment of the contribution of transplanted Scottish Atlantic salmon (*Salmo salar*) to the Esva River (Northern Spain). *Canadian Journal of Fisheries and Aquatic Sciences* **51**, 248-252.

Mork, J. 1991. One-generation effects of farmed fish immigration on the genetic differentiation of wild Atlantic salmon in Norway. *Aquaculture* **98**, 267-276.

Murphy, T. 1980. Studies on precocious maturity in artificially reared Atlantic salmon parr *Salmo salar* L. Ph.D. dissertation. University of Stirling, Scotland.

Myers, R.A. 1986. Game theory and the evolution of Atlantic salmon (*Salmo salar*) age at maturation. In *Salmonid Age at Maturity* (Meerburg, D.J. ed.) *Canadian Special Publications in Fisheries & Aquatic Sciences* **89**, 53-61.

Myers, R.A., Hutchings, J.A. & Gibson, R.J. 1986. Variation in male parr maturation within and among populations of Atlantic salmon, *Salmo salar*. *Canadian Journal of Fisheries & Aquatic Sciences* **43**, 1242-1248.

Nævdal, G. 1983. Genetic factors in connection with age at maturation. *Aquaculture* **33**, 97-106.

Nævdal, G., Leroy, R. & Møller, D. 1983. Sources of variation in weight and length of Atlantic salmon. *Fiskeridirektoratets Skrifter Serie Havundersøkelser* **17**, 359-366.

Nævdal, G., Holm, M., Leroy, R. & Møller, D. 1978. Individual growth rate and age at first maturity in Atlantic salmon. *Fiskeridirektoratets Skrifter Serie Havundersøkelser* **16**, 519-529.

Nævdal, G., Holm, M., Møller, D. & Osthus, O.D. 1975. Experiments with selective breeding of Atlantic salmon. *International Council for the Exploration of the Sea* **CM.1975/M:22**.

Nei, M. 1973. Analysis of gene diversity in subdivided populations. *Proceedings of the National Academy of Sciences of the USA* **70**, 3321-3323.

Nicieza, A.G., Brana, F. & Toledo, M.M. 1991. Development of length-bimodality and smolting in wild stocks of Atlantic salmon, *Salmo salar* L., under different growth conditions. *Journal of Fish Biology* **38**, 509-523.

Nordqvist, O. 1924. Times of entering of the Atlantic salmon (*Salmo salar* L.) in the rivers. *Rapports et Proces-Verbaux, Réunion Conseil Permanent International pour l'Exploration de la Mer* **33**, 1-58.Nygren, A., Nilsson, B. & Jahnke, M. 1972. Cytological studies in Atlantic salmon from Canada, in hybrids between Atlantic salmon from Canada and Sweden, and in hybrids between Atlantic salmon and sea trout. *Hereditas* **70**, 295-306.

Nyman, L. 1967. Protein variations in Salmonidae. *Report of the Institute of Freshwater Research (Drottningholm)* **47**, 5-38.

Nyman, L. & Norman, L. 1987. *Genetic aspects on culture of Atlantic salmon and sea trout for stocking: Guidelines for breeding methodology and management*. Alvkarleby, Sweden: Salmon Research Institute.

Ohno, S., Wolf, U. & Atkin, N.S. 1968. Evolution from fish to mammals by gene duplication. *Hereditas* **39**, 169-187.

Ovenden,J.R., Bywater, R. & White, R.W.G. 1993. Mitochondrial DNA nucleotide sequence variation in Atlantic salmon (*Salmo salar*), brown trout (*S.trutta*), rainbow trout (*Oncorhynchus mykiss*) and brook trout (*Salvelinus fontinalis*) from Tasmania, Australia. *Aquaculture* **114**, 217-227.

Paloheimo, J.E. & Elson, P.F. 1974. Reduction of Atlantic salmon (*Salmo salar*) catches in Canada attributed to the Greenland fishery. *Journal of the Fisheries Research Board of Canada* **31**, 1467-1480.

Palva, T.K., Lehvaslaiho, H. & Palva, E.T. 1989. Identification of anadromous and non-anadromous salmon stocks in Finland by mitochondrial DNA analysis. *Aquaculture* **81**, 237-244.

Payne, R.H. 1980. The use of serum transferrin polymorphism to determine the stock composition of Atlantic salmon in the west Greenland fishery. *Rapports et Proces-Verbaux, Réunion Conseil Permanent International pour l'Exploration de la Mer* **176**, 60-64.

Peterson, R.H., Spinney, H.C.E. & Sreedharan, A. 1977. Development of Atlantic salmon (*Salmo salar*) eggs and alevins under varied temperature regimes. *Journal of the Fisheries Research Board of Canada* **34**, 31-43.

Piggins, D.J. 1980. Salmon ranching in Ireland. In *Salmon Ranching*(Thorpe, J.E. ed.) pp.187-198. London: Academic Press.

Piggins, D.J. 1983. Census work on fish movements: Reared salmon: summary of selective breeding programme. In Annual Report, Salmon Research Trust of Ireland **27**.

Pope, J.A., Mills, D.H. & Shearer, W.M. 1961. The fecundity of Atlantic salmon (*Salmo salar* L.). *Freshwater Salmon Fisheries Research* **26**,1-12.

Porter, T.R., Healey, M.C., O'Connell, F., Baum, E.T., Bielak, A.T. & Côté, Y. 1986. Implications of varying the sea age at maturity of Atlantic salmon (*Salmo salar*) on yield to the fisheries. In *Salmonid Age at Maturity* (Meerburg, D.J. ed.) *Canadian Special Publications in Fisheries & Aquatic Sciences* **89**, 110-117.

Power, G. 1981. Stock characteristics and catches of Atlantic salmon (*Salmo salar*) in Quebec, and Newfoundland and Labrador in relation to environmental variables. *Canadian Journal of Fisheries & Aquatic Sciences* **38**, 1601-1611.

Reddin, D.G. & Misra, R.K. 1985. Hotelling's T2 to identify the origin of Atlantic salmon (*Salmo salar*) in a mixed-stock fishery. *Canadian Journal of Fisheries & Aquatic Sciences* **42**, 250-255.

Reddin, D.G. & Short, P.B. 1985. Identification of North American and European Atlantic salmon (*Salmo salar* L.) caught at west Greenland in 1984. *International Council for the Exploration of the Sea* **CM.1985/M:12**.

Rees, H. 1967. The chromosomes of *Salmo salar*. *Chromosoma (Berlin)* **21**, 472-474.

Refstie, T. 1986. Genetic differences in stress response in Atlantic salmon and rainbow trout. *Aquaculture* **57**, 374.

Refstie, T. & Steine, T.A. 1978. Selection experiments with salmon III. Genetic and environmental sources of variation in length and weight of Atlantic salmon in the freshwater phase. *Aquaculture* **14**, 221-234.

Refstie, T., Steine, T.A. & Gjedrem, T. 1977. Selection experiments with salmon. II. Proportion of Atlantic salmon smoltifying at 1 year of age. *Aquaculture* **10**, 231-242.

Reimers, E., Kjørrefjord, A.G. & Stavøstrand, S.M. 1993. Compensatory growth and reduced maturation in second sea winter farmed Atlantic salmon following starvation in February and March. *Journal of Fish Biology* **43**, 805-810.

Ricker, W.E. 1972. Hereditary and environmental factors affecting certain salmonid populations. In *The Stock Concept in Pacific Salmon* (Simon, R.C. & Larkin, P.A.eds.) pp.19-160. H.R.Macmillan Lectures in Fisheries, Vancouver: University of British Columbia.

Riddell, B.E. 1986. Assessment of selective fishing on the age at maturity in Atlantic salmon (*Salmo salar*): A genetic perspective. In *Salmonid Age at Maturity* (Meerburg, D.J. ed.) *Canadian Special Publications in Fisheries & Aquatic Sciences* **89**, 102-109.

Riddell, B.E. & Leggett, W.C. 1981. Evidence of an adaptive basis for geographic variation in body morphology and time of downstream migration of juvenile Atlantic salmon (*Salmo salar*). *Canadian Journal of Fisheries & Aquatic Sciences* **38**, 308-320.

Riddell, B.E., Leggett, W.C. & Saunders, R.L. 1981. Evidence of adaptive polygenic variation between two populations of Atlantic salmon (*Salmo salar*) native to tributaries of the S.W. Miramichi River, N.B. *Canadian Journal of Fisheries & Aquatic Sciences* **38**, 321-333.

Ritter, J.A. 1975. Lower ocean survival rates for hatchery-reared Atlantic salmon (*Salmo salar*) stocks released in rivers other than native streams. *International Council for the Exploration of the Sea* **CM.1975/M:26**

Ritter, J.A. & Newbould, K. 1977. Relationship of parentage and smolt age to age at first maturity of Atlantic salmon (*Salmo salar*). *International Council for the Exploration of the Sea* **CM.1977/M:32.**

Ritter, J.A., Farmer, G.J., Misra, R.K., Goff, T.R., Bailey, J.K & Baum, E.T. 1986. Parental influences and smolt size and sex ratio effects on sea age at first maturity of Atlantic salmon (*Salmo salar*). In *Salmonid Age at Maturity* (Meerburg, D.J. ed.) *Canadian Special Publications in Fisheries & Aquatic Sciences* **89**, 30-38.

Roberts, F.L. 1970. Atlantic salmon (*Salmo salar*) chromosomes and speciation. *Transactions of the American Fisheries Society* **99**, 105-111.

Ros, T. 1981. Salmonids in the Lake Vänern area. *Ecological Bulletin (Stockholm)* **34**, 21-31.Rosenau, M.L., & McPhail, J.D. 1987. Inherited differences in agonistic behaviour between two populations of Coho salmon. *Transactions of the American Fisheries Society* **116**, 646-654.

Rowe, D.K. & Thorpe, J.E. 1990a. Differences in growth between maturing and non-maturing male Atlantic salmon (*Salmo salar* L.) parr. *Journal of Fish Biology* **36**, 643-658.

Rowe, D.K. & Thorpe, J.E. 1990b. Suppression of maturation in male Atlantic salmon (*Salmo salar* L.) parr by reduction in feeding and growth during spring months. *Aquaculture* **86**, 291-313.

Rowe, D.K., Thorpe, J.E. & Shanks, A.M. 1991. Role of fat stores in the maturation of male Atlantic salmon (*Salmo salar*) parr. *Canadian Journal of Fisheries & Aquatic Sciences* **48**, 405-413.

Ryman, N. 1970. A genetic analysis of recapture frequencies of released young salmon.*Hereditas* **65,** 159-160.

Ryman, N. 1981. Conservation of genetic resources: experiences from the brown trout (*Salmo trutta*). *Ecological Bulletin (Stockholm)* **34**, 61-74.

Ryman, N. 1983. Patterns of distribution of biochemical genetic variation in salmonids: differences between species. *Aquaculture* **33**, 1-21.

Sadler, S.E., Friars, G.W. & Ihssen, P.E. 1986. The influence of temperature and genotype on the growth rate of hatchery-released salmonids. *Canadian Journal of Animal Science* **66**, 599-606.

Saunders, R.L. 1981. Atlantic salmon (*Salmo salar*) stocks and management implications in the Canadian Atlantic provinces and New England, USA. *Canadian Journal of Fisheries & Aquatic Sciences* **38,** 1612-1625.

Saunders, R.L. 1986. The scientific and management implications of age and size at sexual maturity in Atlantic salmon (*Salmo salar*). In *Salmonid Age at Maturity* (Meerburg, D.J. ed.) *Canadian Special Publications in Fisheries & Aquatic Sciences* **89**, 3-6.

Saunders, R.L. 1991. Potential interaction between cultured and wild Atlantic salmon. *Aquaculture* **98**, 51-60.

Saunders, R.L. & Schom, C.B. 1985. Importance of the variation in life history parameters of Atlantic salmon (*Salmo salar*). *Canadian Journal of Fisheries & Aquatic Sciences* **42**, 615-618.

Saunders, R.L. & Sreedharan, A. 1977. The incidence and genetic implications of sexual maturity in male Atlantic salmon parr. *International Council for the Exploration of the Sea* **CM.1977/M:21.**

Saunders, R.L., Henderson, E.B. & Glebe, B.D. 1982. Precocious sexual maturation and smoltification in male Atlantic salmon (*Salmo salar*). *Aquaculture* **28**, 211-229.

Saunders, R.L., Henderson, E.B., Glebe, B.D. & Loudenslager, E.J. 1983. Evidence of a major environmental component in determination of the grilse: Larger salmon ratio in Atlantic salmon (*Salmo salar*). *Aquaculture* **33**, 107-118.

Saunders, R.L., Benfey, T., Bradley, T.M., Duston,J., Farmer, G., McCormick, S.D. & Specker, J.L. (eds.).1994. Salmonid Smolting Workshop IV. *Aquaculture* **121**, 1-300.

Scarnecchia, D.L. 1983. Age at sexual maturity in Icelandic stocks of Atlantic salmon (*Salmo salar*). *Canadian Journal of Fisheries & Aquatic Sciences* **40**,1456-1468.

Schaffer, W.M. 1974. Selection for optimal life histories: The effect of age structure. *Ecology* **55**, 291-303.

Schaffer, W.M. & Elson, P.F. 1975. The adaptive significance of variations in life history among local populations of Atlantic salmon in North America. *Ecology* **56**, 577-590.

Schom, C.B. 1986. Genetic, environmental, and maturational effects on Atlantic salmon (*Salmo salar*) survival in acute low pH trials. *Canadian Journal of Fisheries & Aquatic Sciences* **43**, 1547-1555.

Schom, C.B. & Davidson, L.A. 1982. Genetic control of pH resistance in Atlantic salmon (*Salmo salar*). *Canadian Journal of Genetics & Cytology* **24**, 636.

Schom, C.B. & Saulnier, C.E. 1983. The relationship between Atlantic salmon family gill morphologic differences and the mean survival time in low pH environment. *Genetics Society of Canada Bulletin* **14**, 42-43.

Scott, A.P. & Baynes, S.M. 1980. A review of the biology, handling and storage of salmonid spermatozoa. *Journal of Fish Biology* **17**, 707-739.

Selander, R.K. 1976. Genic variation in natural populations. In *Molecular Evolution* (Ayala, F.J. ed.) pp.21-45. Sunderland, Mass.: Sinauer Associates.

Simon, R.C. & Larkin, P.A. 1972. *The Stock Concept in Pacific Salmon.* H.R.MacMillan Lectures in Fisheries. Vancouver, Canada: University of British Columbia.

Slatkin, M. 1987. Gene flow and the geographic structure of natural populations. *Science* **236**, 787-792.

Slynko, V.I., Semenova, S.K. & Kazakov, R.V. 1981. Izucheniye populyatsionno-geneticheskoyi struktury atlanticheskogo lososya v svyazi s zadachami ego razvedeniya. Soobshch.II. Analiz chastot fenotipov malikenzima v populyatsiyakh lososya rek Nevy i Norovy. *Sbornik nauchnykh trudov GosNIORKh* **163**, 124-128.

Stabell, O.B. 1984. Homing and olfaction in salmonids: A critical review with special reference to the Atlantic salmon. *Biological Reviews* **59**, 333-388.

Ståhl, G. 1981. Genetic differentiation among natural populations of Atlantic salmon (*Salmo salar*) in northern Sweden. *Ecological Bulletin (Stockholm)* **34**, 95-105.

Ståhl, G. 1983. Differences in the amount and distribution of genetic variation between natural populations and hatchery stocks of Atlantic salmon. *Aquaculture* **33**, 23-32.

Anadromous and resident Atlantic salmon spawners: two contrasting life-history strategies – large migrant female (c.5 years old, c.4 kg) and small non-migrant male (c.2 years old, c.20 g). (Photograph courtesy of David Hay.) The size discrepancy between this mature adult male, and the 18 kg one on the cover photograph is even more remarkable.

Washington Shellfish Farmers produce a variety of species and sizes of oysters.

Washington State Shellfish Farmers use "Off Bottom" Culture methods to produce oysters for the half-shell market.

(Photographs courtesy of P. Davis, Pacific Coast Oyster Growers Association.)

Genetically distinct forms of common carp.
(1) Gold – a European inbred type. (2) Blue-grey – a European inbred type. (3) Wild – a European crossbred type. (4) Wild – Chinese big-belly carp. (5) Gold × Chinese – an interracial crossbred type. (Photographs courtesy Dr Gideon Hulata.)

Philippine (above) and GIFT Program selected (below) Nile tilapia. (Photograph courtesy of Dr Ambekar Eknath.)

Ståhl, G. 1987. Genetic population structure of Atlantic salmon. In *Population Genetics and Fishery Management* (Ryman, N. & Utter, F. eds.) pp.121-140. Seattle: University of Washington Press.

Ståhl, G., Loudenslager, E.J., Saunders, R.L. & Schofield, E.J. 1983. Electrophoretic study on Atlantic salmon populations from the Miramichi River (New Brunswick) System, Canada. *International Council for the Exploration of the Sea* **CM.1983/M:20.**

Stearns, S.C. & Crandall, R.E. 1984. Plasticity for age and size at maturity: a life-history response to unavoidable stress. In *Fish Reproduction: Strategies and Tactics* (Potts, G.W. & Wootton, R.J. eds.) pp.13-33. London: Academic Press.

Stock Concept International Symposium. 1981. Stock concept international symposium.*Canadian Journal of Fisheries & Aquatic Sciences* **38**, 1457-1921.

Stoss, J. & Refstie, T. 1982. Short term storage and cryopreservation of milt from Atlantic salmon and sea trout. *Aquaculture* **30**, 229-236.

Struthers, G. & Stewart, D. 1986. Observations on the timing of migration of smolts from natural and introduced juvenile salmon in the upper River Tummel, Scotland. *International Council for the Exploration of the Sea* **CM.1986/M:4.**

Sutterlin, A.M. & MacLean, D. 1984. Age at first maturity and the early expression of oocyte recruitment process in two forms of Atlantic salmon (*Salmo salar*) and their hybrids. *Canadian Journal of Fisheries & Aquatic Sciences* **41**, 1139-1149.

Sutterlin, A.M., Henderson, E.B., Merrill, S.P., Saunders, R.L. & MacKay, A.A. 1981. Salmonid rearing trials at Deer Island, New Brunswick, with some projections on economic viability. *Canadian Technichal Report in Fisheries & Aquatic Sciences* **1011**.

Taylor, E.B. 1991. A review of local adaptation in Salmonidae, with particular reference to Pacific and Atlantic salmon. *Aquaculture* **98**, 185-207.

Thorpe, J.E. 1975. Early maturity in male Atlantic salmon. *Scottish Fisheries Bulletin* **42**, 15-17.

Thorpe, J.E. 1977. Bimodal distribution of length of juvenile Atlantic salmon (*Salmo salar*) under artificial rearing conditions. *Journal of Fish Biology* **11**, 175-184.

Thorpe, J.E. 1986. Age at first maturity in Atlantic salmon, *Salmo salar*: Freshwater period influences and conflict with smolting. In *Salmonid Age at Maturity* (Meerburg, D.J. ed.) *Canadian Special Publications in Fisheries & Aquatic Sciences* **89**, 7-14.

Thorpe, J.E. 1987a. Environmental regulation of growth patterns in juvenile Atlantic salmon. In *Age and Growth in Fish* (Summerfelt, R.C. & Hall, G.E. eds.) pp. 463-474. Ames: Iowa State University Press.

Thorpe, J.E. 1987b. Smolting versus residency: Developmental conflict in salmonids. *American Fisheries Society Symposium* **1**, 244-252.

Thorpe, J.E. 1991. Acceleration and deceleration effects of hatchery rearing on salmonid development and their consequences for wild stocks. *Aquaculture* **98**, 111-118.

Thorpe, J.E. 1994a. An alternative view of smolting in salmonids. *Aquaculture* **121**, 105-113.

Thorpe, J.E. 1994b. Reproductive strategies in Atlantic salmon, *Salmo salar* L. *Aquacultre & Fisheries Management* **25**, 77-87.

Thorpe, J.E. 1994c. Impacts of fishing on genetic structure of salmonid populations. In *Genetic Conservation of Salmonid Fishes*, (Cloud, J.G. & Thorgaard, G. eds.) pp.68-81. New York: Plenum.

Thorpe, J.E. 1994d. Significance of straying in salmonids and implications for ranching. *Aquaculture & Fisheries Management* **25**,

Thorpe, J.E. & Mitchell, K.A. 1981. Stocks of Atlantic salmon (*Salmo salar*) in Britain and Ireland: Discreteness and current management. *Canadian Journal of Fisheries & Aquatic Sciences* **38**, 1576-1590.

Thorpe, J.E. & Morgan, R.I.G. 1978. Parental influence on growth rate, smolting rate and survival in hatchery reared juvenile Atlantic salmon, *Salmo salar*. *Journal of Fish Biology* **13**, 549-556.

Thorpe, J.E. & Morgan, R.I.G. 1980. Growth-rate and smolting-rate of progeny of male Atlantic salmon parr, *Salmo salar* L. *Journal of Fish Biology* **17**, 451-459.

Thorpe, J.E., Miles, M.S. & Keay, D.S. 1984. Developmental rate, fecundity and egg size in Atlantic salmon, *Salmo salar* L. *Aquaculture* **43**, 289-305.

Thorpe, J.E., Adams, C.E., Miles, M.S. & Keay, D.S. 1989. Some photoperiod and temperature influences on growth opportunity in juvenile Atlantic salmon, *Salmo salar* L. *Aquaculture* **82**, 119-126.

Thorpe, J.E., Bern, H.A., Saunders, R.L. & Soivio, A. eds. 1985. Salmon Smoltification Workshop II. *Aquaculture* **45**, 1-404.

Thorpe, J.E., Morgan, R.I.G., Talbot, C. & Miles, M.S. 1983. Inheritance of developmental rates in Atlantic salmon, *Salmo salar* L. *Aquaculture* **33**, 119-128.

Thorpe, J.E., Talbot, C., Miles, M.S. & Keay, D.S. 1991. Control of maturation in cultured Atlantic salmon, *Salmo salar*, in pumped seawater tanks, by restricting food intake. *Aquaculture* **86**, 315-326.

Thorpe, J.E. & Koonce, J.F., with Borgeson, D., Henderson, B., Lamsa, A., Maitland, P.S., Ross, M.A., Simon, R.C. & Walters, C. 1981. Assessing and managing man's impact on fish genetic resources. *Canadian Journal of Fisheries & Aquatic Sciences* **38**, 1899-1907.

Torrissen, K.R. 1987. Genetic variation of trypsin-like isozymes correlated to fish size in Atlantic salmon. *Aquaculture* **62**,1-10.

Torrissen, K.R. 1991. Genetic variation in growth rate of Atlantic salmon with different trypsin-like isozyme patterns. *Aquaculture* **93**, 299-312.

Torrissen, K.R. & Shearer, K.D. 1992. Protein digestion, growth and food conversion in Atlantic salmon and Arctic charr with different trypsin-like isozyme patterns. *Journal of Fish Biology* **41**, 409-415.

Torrissen, K.R., Male, R. & Naevdal, G. 1993. Trypsin isozymes in Atlantic salmon,*Salmo salar* L.: studies of heredity, egg quality and effect on growth of three different populations. *Aquaculture & Fisheries Management* **24**, 407-415.

Verspoor, E. 1986. Spatial correlations of transferrin allele frequencies in Atlantic salmon (*Salmo salar*) populations from North America. *Canadian Journal of Fisheries & Aquatic Sciences* **43**, 1074-1078.

Verspoor, E. 1988. Reduced genetic variability in first generation hatchery populations of Atlantic salmon (*Salmo salar*). *Canadian Journal of Fisheries & Aquatic Sciences* **45**, 1686-1690.

Verspoor, E. & Cole, L.J. 1989. Genetically distinct sympatric populations of resident and anadromous Atlantic salmon, *Salmo salar* L. *Canadian Journal of Zoology* **67**, 1453-1462.

Verspoor, E. & Jordan, W.C. 1989. Genetic variation at the Me-2 locus in the Atlantic salmon within and between rivers: Evidence for its selective maintenance. *Journal of Fish Biology* **35 (Supplement A)**, 205-213.

Verspoor, E., Fraser, N.H.C & Youngson, A.F. 1991. Protein polymorphism in Atlantic salmon within a Scottish river: Evidence for selection and estimates of gene flow between tributaries. *Aquaculture* **98**, 217-230.

Villarreal, C.A., Thorpe, J.E. & Miles, M.S. 1988. Influence of photoperiod on growth changes in juvenile Atlantic salmon, *Salmo salar* L. *Journal of Fish Biology* **33**, 15-30.

Vuorinen, J. 1982. Little genetic variation in Finnish lake salmon, *Salmo salar sebago* (Girard). *Hereditas* **97**, 189-192

Vuorinen, J. & Berg, O.K. 1989. Genetic divergence of anadromous and non-anadromous Atlantic salmon (*Salmo salar*) in the River Namsen, Norway.*Journal of Fish Biology* **46**, 406-409.

Waples, R.S. 1991. Genetic interactions between hatchery and wild salmonids: lessons from the Pacific Northwest. *Canadian Journal of Fisheries & Aquatic Sciences* **48** (**Supplement 1**), 124-133.

Waples, R.S. & Teel, D.J. 1990. Conservation genetics of Pacific salmon. I. Temporal changes in allele frequency. *Conservation Biology* **4**, 144-156.

Waples, R.S.,Winans, G.A., Utter, F.M. & Mahnken, C. 1990. Genetic approaches to the management of Pacific Salmon. *Fisheries* **15**, 19-25.

Ward, R.D. & Beardmore, J.A. 1977. Protein variation in the plaice, *Pleuronectes platessa* L. *Genetics Research* **30**, 45-62.

Webb, J.H., Hay, D.W., Cunningham, P.D. & Youngson, A.F. 1991. The spawning behaviour of escaped farmed and wild adult Atlantic salmon (*Salmo salar* L.) in a northern Scottish river. *Aquaculture* **98**, 97-110.

Went, A.E.J. 1969. Irish kelt tagging experiments, 1961/62 to 1966/67. *Irish Fisheries Investigations Series A* **5**, 34-47.

Wohlfarth, G.W. 1986. Decline in natural fisheries - a genetic analysis and suggestion for recovery. *Canadian Journal of Fisheries & Aquatic Sciences* **43**, 1298-1306.

Youngson, A.F., Martin, S.A.M., Jordan, W.C. & Verspoor, E. 1991. Genetic protein variation in Atlantic salmon in Scotland: Comparison of wild and farmed fish. *Aquaculture* **98**, 231-242.

Zelinskiy, Yu.P. & Medvedeva, I.M. 1985. Analysis of chromosomal variability and polymorphism in Atlantic salmon, *Salmo salar*, of Lake Onega. *Journal of Ichthyology* **25**, 70-77.

6
THE CUPPED OYSTER AND THE PACIFIC OYSTER

D.Hedgecock

Oysters are good examples of benthic marine species upon which concern for the conservation of aquatic genetic resources ought to be focused. They are pervasively distributed throughout the world's near-shore, shallow water, bay, and estuarine habitats between 64°N and 44°S, where they serve as indicator species of environmental change and quality (Beattie *et al.*, 1982; Galtsoff, 1964; Jackson, 1988; Yonge, 1960). Moreover, they have been widely harvested by humans since prehistoric times, as shell middens in Europe, North America, and Asia attest.

Today, most of the world's oyster production is from intensive culture or enhanced oyster grounds (Bourne, 1986). Cupped oysters (genus *Crassostrea*) have become more important than their relatives, the larviparous flat oysters (genus *Ostrea*), whose native stocks have been devastated by overharvesting and epidemic diseases in Europe (Clark, 1959; Grizel *et al.*, 1988) and elsewhere (Barrett, 1963). Of the nearly 1 million metric tons of oysters landed in the world in 1980 (4.2% of total world fishery production), 98% were cupped oysters and only 2% were flat oysters. Of total landings, 53% consisted of a single species, the Japanese or Pacific cupped oyster *Crassostrea gigas* (statistics from the Food and Agriculture Organization of the United Nations reported in Bourne, 1986). For cupped oysters, propagation of seed by hatcheries and complete control over the life cycle are becoming increasingly important (Chew, 1984). The species advancing most rapidly towards domestication is the Pacific oyster *C. gigas* (Hedgecock, 1988; Hershberger *et al.*, 1984).

COMPONENTS OF DIVERSITY

Mode of Reproduction

Cupped oysters of the genus *Crassostrea* are among the atypical metazoan animals in which truly random mating, that which allows for self-fertilization, is possible (Galtsoff, 1964). The reproductive mode of these oysters is protandric, sequential, irregularly alternating hermaphroditism, that is, they tend to mature in their first reproductive season as males, they often change in their second reproductive season into females, and thereafter they may alternate sex depending on environmental circumstances. Individuals that are functionally hermaphroditic and self-compatible occur typically at frequencies of a few per thousand.

Sex determination appears to be highly labile, so that a given individual need not follow the typical pattern described. Genetic control of sex is thought to reside in a small number of autosomal genes, each having male- and female-determining alleles (Haley, 1977, 1979). The sex of an individual is determined by the balance among these alleles, and the sex ratio and frequency of sex change in a population of oysters is determined in part by the distribution of multilocus genotypes and in part by the environment.

Female cupped oysters are oviparous and shed small eggs (40μm in diameter) into the sea usually after the males have spawned in response to rising temperatures (Galtsoff, 1964; Quayle, 1969). The remarkable fecundity of oysters is well known. An individual female oyster spawns up to 100 million eggs at a time, probably more than once per season. Yet fecundity is highly variable, depending primarily on body size, food availability, temperature during the conditioning phase, and genotype (Lannan, 1980a,b; Lannan *et al.*, 1980). Males generally produce even more copious numbers of gametes, but the gametes' competency to fertilize eggs can vary dramatically (Lannan, 1980a,b).

Larval Development and Early Mortality

Within 5 to 10 hours after fertilization, the oyster egg develops into a swimming, nonfeeding, trochophore larva, which 48 hours after fertilization develops into a planktotrophic, straight-hinge, veliger larva (Galtsoff, 1964). The first two to three weeks of an oyster's life are then spent as a free-swimming, pelagic larva whose dispersal away from its parents is virtually guaranteed by tidal and oceanic currents far stronger than its feeble swimming abilities. Larval dispersal implies high rates of gene flow among geographic populations (Buroker, 1983, 1984; Grady *et al.*, 1984), but the adaptive and evolutionary significance of this exchange needs to be carefully evaluated (Hedgecock, 1986).

Major anatomical changes in the larva, particularly the development of a foot and a pair of pigmented eye spots, herald the approach of metamorphosis and settlement (Galtsoff, 1964). Selection of an appropriate setting site, cementation of the left shell to the substrate, and metamorphosis, processes under endogenous and exogenous chemical control (Bonar *et al.*, 1990; Coon *et al.*, 1986), terminate larval life. Metamorphosis itself involves dramatic anatomical reorganizations: losses of the velum and foot, migrations and rearrangements of major organ systems, and enlargements of gill and adductor muscle tissues. These changes take place during a 24-hour period during which the larva cannot feed. What results is a benthic, filter-feeding, juvenile oyster that grows into an adult within a few years.

Mortality of larvae and newly metamorphosed juveniles (spat) is a significant feature of oyster life history. For the larviparous European flat oyster *Ostrea edulis*, investigators estimate that 1 million larvae may yield only 250 attached spat, of which 95% perish before the onset of winter (Clark, 1959; Galtsoff, 1964). Rates of survival for the planktotrophic larvae of *Crassostrea* are likely to be at least an order of magnitude lower. Survival of naturally spawned *C. gigas* larvae in Pendrell Sound, British Columbia, varies from 0-9% for the straight-hinge to eyed larval period (Quayle, 1969). To this mortality must be added the mortality of newly settled spat, which has not been estimated in nature. Under artificial culture, 5-10% of fertilized *C. gigas* eggs typically survive to metamorphosis, of which only up to 25% are recovered as spat (Lannan, 1980a). Researchers believe that nutrition and reproductive conditioning of broodstock, as well as the dispersal, predation, and nutrition of the larvae themselves, play important roles in this early mortality. Genetic factors that have been implicated are discussed later.

Adult Population Biology

The distribution and abundance of adult oyster populations have been recorded throughout history (Clark, 1959). Indeed, oyster reefs once posed navigational hazards in bays like the Chesapeake Bay (Jackson, 1988). Although the chief causes of most historical and modern population declines appear to be anthropogenic, some variation in abundance may occur naturally. Density-independent factors influence variation in larval recruitment for most marine organisms, but remain rather poorly understood for benthic invertebrates (Dayton, 1979; Gaines & Roughgarden, 1985; Gaines *et al.*, 1985; Roughgarden *et al.*, 1988; Underwood & Denley, 1984).

Bivalve populations frequently comprise one or two strong-year classes, owing to sporadically successful recruitment, for example, Pacific oysters in the Strait of Georgia (Bourne, 1986; Quayle, 1969). In addition, native oyster populations are periodically decimated by epizootic diseases, parasitic infestations, or outbreaks of predator populations (Beattie *et al.*, 1988; Clark, 1959; Galtsoff, 1964; Grizel *et al.*, 1988; Korringa, 1976). Whether such disasters constitute bottlenecks with respect to the transmission of genetic diversity is unknown.

Galtsoff (1964) lists five factors that favour the development of oyster communities - firm bottom character, water movement, salinity, temperature, and food - and five factors that inhibit their development: sedimentation, pollution, competition, disease, and predation. Under unfavourable conditions other dominant bivalves, such as mussels or cockles, can replace oyster communities rapidly (Clark, 1959; Galtsoff, 1964). Socioeconomic factors, such as changing species preference, can significantly affect rates of human predation and bivalve community structure (Bourne, 1986).

Oyster populations are naturally subdivided among beds or reefs within bays or estuaries, among discrete bay or estuarine systems along continental margins, or among islands. Do these geographic populations represent genetically distinct subdivisions as well? Dispersal of pelagic larvae is widely thought to homogenize the gene pools of most species of oyster, but conflicting pictures of genetic subdivision emerge from studies at different biological levels (Hedgecock & Berthelemy-Okazaki, 1984). Electrophoretic studies of protein variation show differentiation only over very broad regions of the order of thousands of kilometres. However, a variety of physiological processes, activities, and tolerances, and biochemical and serological traits vary on spatial scales of tens to hundreds of kilometres (Hillman, 1964, 1965; Li *et al.*, 1967; Menzel, 1951, 1956; Numachi, 1962; Stauber, 1947, 1950). The maintenance of geographic differences in spawning cycles of oysters even after transplantation and acclimation to a distant site is strong evidence for adaptive divergence of reproductive characters (Loosanoff & Nomejko, 1951). The existence of physiological races implies that local broodstock should be used in restocking or enhancement efforts.

In contrast to the evidence for inherent physiological differences among oysters from different geographic populations, variation in shell morphology, which can be quite striking, is largely determined by local environmental conditions and shows no consistent geographic pattern of variation (Galtsoff, 1964). At the chromosome level, oysters are surprisingly uniform. Karyotypes of members of the family Ostreidae are quite similar and have a haploid number of 10 (Menzel, 1968).

Genetic Diversity Within and Between Cupped Oyster Populations

Application of protein electrophoresis to the estimation of genetic diversity in cupped oysters has shown them to be among the more genetically variable animals known (Table 6.1). Typically half of the enzymes and proteins surveyed in oysters prove to be polymorphic within local populations, and an index of genetic diversity, the average expected proportion of heterozygous loci per individual, ranges from 0.07-0.24, with an arithmetic average over 16 species of 0.17. For comparison, average heterozygosity over 551 vertebrate species that have been assayed by similar methods is 0.054 ± 0.003 (Nevo *et al.*, 1984).

Another striking feature of oyster population genetics is a consistent deficiency of observed proportions of heterozygotes with respect to proportions expected under the assumption of random mating or Hardy-Weinberg-Castle equilibrium (reviewed by Zouros & Foltz, 1987), measured by D in Table 6.1. Though widespread among bivalve molluscs, such deficiencies of heterozygotes (which may also be viewed as excesses of homozygotes) are not typical of animal populations, not even of other benthic marine organisms with comparable life histories, for example, decapod crustaceans (Hedgecock *et al.*, 1982). Heterozygote deficiency in bivalve molluscs is one of the most prominent, unexplained phenomena in population genetics.

The handful of studies of quantitative genetic variation in oysters have all revealed substantial components of additive genetic variance (Table 6.2). Most studies have been concerned only with larval traits, but a few have dealt with adult characteristics (Gjedrem, 1983). Researchers

Table 6.1. Genetic diversity within 16 species of the cupped oysters *Crassostrea* and *Saccostrea* as estimated by protein electrophoresis.

Species	Proportion of loci polymorphic	Heterozygosity Observed	Heterozygosity Expected*	D†
Crassostrea angulata	0.60	0.234	0.238	-0.017
C.belcheri	0.20	0.062	0.068	-0.088
C.corteziensis	0.65	0.138	0.196	-0.296
C.gigas	0.53	0.195	0.216	-0.097
C.irredalei	0.39	0.100	0.105	-0.048
C.nippona	0.42	0.098	0.109	-0.101
C.rhizophorae	0.45	0.115	0.131	0.122
C.rivularis	0.37	0.095	0.118	-0.195
C.sikamea	0.60	0.189	0.210	-0.100
C.virginica	0.54	0.219	0.215	+0.019
C. sp.	0.53	0.096	0.097	-0.007
Averages:	0.48	0.140	0.155	-0.096
Saccostrea commercialist	0.46	0.180	0.190	-0.053
S.cucculata	0.48	0.185	0.180	+0.028
S.glomerata	0.49	0.182	0.199	-0.085
S.malabonensis	0.53	0.192	0.213	-0.099
S.manilai	0.47	0.187	0.194	-0.036
Averages:	0.49	0.185	0.195	-0.049
Grand averages:	0.48	0.154	0.167	-0.081

* Expected heterozygosity is the average of values for each locus calculated as $1-\sum p_i^2$ where p_i are the frequencies of alleles ($\sum p_i = 1$).
† $D = (H_o-H_e)/H_e$. A negative value indicates a deficiency of the observed with respect to the expected numbers of heterozygous genotypes.

[Compiled from a variety of sources, principally Okazi & Fujio (1985).]

have tended to focus on growth rate and larval survival (e.g., Lannan, 1980a). Lannan's (1980a,b; Lannan *et al.*, 1980) work on components of variance in larval survival has interesting implications for the genetic structure of natural populations. There appears to be substantial additive genetic variance in rates of reproductive maturation in Pacific oysters. A strong seasonal component to the success of hatchery conditioning suggests that different segments of a natural population may be sexually ripe and spawnable at different times. To the extent that such spawning propensities are genetically determined, assortative mating may reduce effective population sizes and genetic diversity.

Table 6.2. General and specific combining abilities as a percentage of total variance and heritabilities (h^2 based on sire S, or family F, components of variance) for quantitative traits in oysters. (Adapted from Gjedrem (1983), and Lannan (1980a).)

Traits	X	General combining ability (%)	Specific combining ability (%)	Number of families oe inbred lines	
Larval survival (%)	35			0.31±0.06	11
Larval growth			0.24		
6-day growth			0.33	0.43	8
17-day length			0.44±0.21		8
16-day growth			0.50	0.60	8
Setting success (%)	45			0.09±0.08	11
6-weeks length			0.50±0.30		5
18-mo meat wt.(g)	30			0.33±0.19	11
18-mo wt (g)	63			0.37±0.06	11
24-mo female gonad index		96.7	0.5		7
24-mo male gonad index		65.4	16.0		7

X = mean for trait.

As already indicated, several studies have failed to demonstrate substantial allozyme frequency variation at spatial scales of less than hundreds to thousands of kilometres for the American oyster *Crassostrea virginica* (Buroker, 1983, 1984; Buroker *et al.*, 1979; Schaal & Anderson, 1974). In general these results support the presumption of substantial gene flow via pelagically dispersing larvae; but at the same time they conflict with evidence for physiological races. Investigators have interpreted slight but statistically significant differences in allelic frequencies among estuaries as the result of local adaptation (Grady *et al.*, 1984; Rose, 1984), but alternative explanations have not been rigorously refuted (Hedgecock, 1982).

On a larger spatial scale, several researchers have described four electrophoretically distinct geographical populations of the American oyster: the Canadian, US Atlantic Coast, US Gulf Coast, and Bay of Campeche populations (Buroker, 1983; Buroker *et al.*, 1979; Hedgecock & Berthemlemy-Okazaki, 1984), with an unusual population in the Laguna Madre, Texas (Groue & Lester, 1982). Reeb & Avise (1990) showed that populations of the American oyster cluster into two major phylogenetic groups, Gulf of Mexico and Atlantic, on the basis of a discontinuity in the distribution of restriction fragment length polymorphisms (RFLPs) of mitochondrial DNA. These two groups are estimated to have diverged about 1.2 million years ago.

Crassostrea gigas from four prefectures in Japan - Hokkaido, Miyagi, Hiroshima, and Kumamoto - were originally classified as races on the basis of morphological, physiological, and reproductive characteristics (Imai &Sakai, 1961). The Kumamoto oyster is reproductively isolated from the other types of Pacific oysters (Imai & Sakai, 1961; Numachi, 1971) and Ahmed (1975) considered it to be a separate species, *C. sikamea*. Early electrophoretic studies of protein variation (Buroker *et al.*, 1979; Fujio, 1979) supported this classification, especially the distinction between Kumamoto oysters and the remaining types. However, more recent electrophoretic evidence indicates that oysters from throughout the Japanese archipelago are genetically very similar (Ozaki & Fujio, 1985). This homogeneity may be due to massive transportation of Miyagi oysters in recent years.

Summary of Inherent Risks to Conservation of Genetic Diversity

For oysters, the Achilles heel of population stability is early life history mortality, which is naturally quite high and variable. Spatfalls of Pacific oysters in Dabob and Quilcene bays in Washington State, for example, provide commercially useful numbers of seed in about 7 of every 10 years (Chew, 1984), and variation in larval density from year to year in the Strait of Georgia has already been mentioned (Quayle, 1969). Yet the causes, and therefore the predictability, of variation in recruitment are undetermined.

Any factor that diminishes successful recruitment to the benthic juvenile phase is likely to place the oyster at substantial risk of population decline. Short-term variation or long-term trends in ocean temperatures, for example, may affect not only the reproductive cycle of adults, but also the growth and development rates of larvae. Changes in oceanic currents could also reduce the probability that larvae find suitable habitats for settlement. Humans may exacerbate early spat mortality in a number of ways. Increased sedimentation in estuaries as the result of upland construction and erosion, for example, can interfere with oyster settlement (Galtsoff, 1964).

Although oyster populations appear to be numerically strong and to have a number of properties that should generate or maintain high levels of genetic diversity, risks to the maintenance of large, effective population sizes can be envisioned. First, oyster populations are distributed over discrete habitat patches, estuaries, and bays. To the extent that geographic subdivision coincides with physiological race formation, loss of a habitat patch may mean the loss of a unique genetic entity. Second, sex ratios are unstable, and may be skewed by environmental perturbation or by indirect selection for some linked or correlated trait. Finally, an assortative mating system based on differences in reproductive responses to environmental factors such as temperature could leave a population particularly vulnerable to reduction in effective size by changes in environment.

Oyster habitats are particularly vulnerable to degradation from human activities and environmental pollution. More insidious may be the effects of overharvesting of natural resources and replacement of these by hatchery-propagated seed, a scenario of increasing importance in oyster production worldwide (Bourne, 1986).

GENETIC PROCESSES

Factors Promoting Genetic Diversity

A number of factors may have contributed to the genetic diversity of unexploited, natural populations of oysters:

- high densities of local populations, thus large effective population sizes;
- high fecundities, thus the possibility for increased mutation or segregation load;
- moderately long and overlapping generations; and
- presumed high rates of gene exchange among geographically separated populations owing to dispersal of pelagic larvae.

Each of these factors needs to be examined critically, especially in exploited populations, to evaluate precisely its role in promoting genetic diversity. To what extent are large local populations actually subdivided into smaller reproductive units by differential mating propensities? Might high mortality in early life history reflect a large variance in the reproductive successes of individuals? Do larvae disperse among estuarine systems effectively? Do they survive to reproduce or to contribute genes to their new population? Unfortunately, few data are available to answer such questions.

Various forms of balancing natural selection can promote high levels of genetic diversity. Researchers have reported a striking correlation between allozyme heterozygosity and fitness-related traits, such as growth rate, for several bivalve species (Koehn, 1984; Mitton & Grant, 1984; Zouros & Foltz 1987). The causes of this correlation are currently controversial (see the exchange between Koehn [1990] and Zouros [1990]). Some maintain that the effect is due to greater fitness of heterozygotes at the allozyme loci in question, although the biochemical or physiological mechanisms of the overdominance are as yet unclear. If true, balancing natural selection would promote heterozygosity at these loci. Others argue that the fitness-heterozygosity correlation is due to associative overdominance, the close linkage of allozyme markers to recessive lethal alleles. Conflicting data on inbreeding depression in larval survival and early growth suggest, however, that the mutational load of recessive lethals may not be large in bivalves (Lannan, 1980b; Longwell & Stiles, 1973; Mallet & Haley, 1983; Mallet *et a*l., 1985).

Factors Restricting Genetic Diversity

Random genetic drift in finite populations erodes genetic diversity. Though drift in oyster populations may seem unlikely, owing to their apparently large sizes, extrinsic and intrinsic restrictions on the effective sizes of oyster populations can be identified. Extrinsic factors reducing oyster populations may be occurring naturally, such as the disease epidemics already cited or anthropogenic perturbations. Here we consider intrinsic population traits or genetic processes that may restrict effective population sizes.

An unstable sex-determination system in oysters (Galtsoff, 1964; Haley, 1977, 1979) allows sex ratios to deviate from the ideal 1:1. In a population with nonoverlapping generations, unequal sex ratios reduce the effective breeding number (N_e) according to the following equation (Wright, 1931, 1969):

$$N_e = 4N_mN_f/(N_m + N_f)$$

where N_m and N_f are the numbers of males and females, respectively, contributing to the gamete pool each generation. For the more complicated case of overlapping generations and different generation times in the two sexes, which applies to oysters, the effective population size per average generation is

$$N_e = 4(T/N_{em}T_m + T/N_{ef}T_f)^{-1}$$

where T_m and T_f are the average ages of reproduction for the two sexes, discounting progeny produced at each age by the growth rate of the population; T is the mean of these two ages; and N_{em} and N_{ef} are the effective numbers of males and females per generation, respectively (Lande & Barrowclough, 1987). Unfortunately, age-specific schedules of fecundity and survival have not been estimated for natural populations of oysters. Galtsoff (1964), summarizing data taken by Coe (1934) for the American oyster, reports an average ratio of males to females in the first breeding season of 11:1. The sex ratio tends to even out with age. Galtsoff (1961) induced spawning in a group of 4-year old American oysters every summer for 5 years. The sex ratio in this group at age 5 was 1.9:1 (N = 182), but by age 9 the sex ratio of 33 survivors was 0.8:1, perhaps owing to greater longevity or survival of females.

A more potent influence on effective population sizes, however, may be variance in reproductive success, which has already been mentioned. Oysters are clearly fecund enough that a relatively small number can repopulate an area. For populations in demographic equilibrium, Wright (1969) provides the relevant relationships among variance in offspring number (V_k), the breeding population number (N), and the effective population number (N_e):

$$N_e = (4N - 4)/(V_k + 2)$$

Thus, variance in the number of offspring that an adult contributes to the breeding population of the next generation is inversely proportional to the effective population number. Unfortunately, no data on variance in offspring number in natural populations are available, but indirect estimates from genetic data of effective population sizes for several marine invertebrates appear to be orders of magnitude less than abundance (Hedgecock *et al.*, 1982; Palumbi & Wilson, 1990), a discrepancy best explained by large V_k. Interestingly, Reeb & Avise (1990) estimate effective female population sizes for the Gulf and Atlantic populations of the American oyster to be only 125,000 and 70,000, respectively. A recent study of hatchery populations of Pacific oysters also implies that variance in reproductive contribution may result in small effective population sizes under artificial propagation (Hedgecock & Sly, 1990).

Finally, we must consider the implications of observed

heterozygote deficiencies in bivalve populations on the maintenance of genetic diversity or effective population sizes. Alternative, but not mutually exclusive explanations for this phenomenon are:

- selection against heterozygotes during embryonic or larval stages (Blanc & Bonhomme, 1987; Singh & Green, 1984);
- admixture of genetically differentiated populations (Wahlund Effect);
- presence of null alleles, aneuploidy, or other biases against recognizing heterozygous allozyme phenotypes (Foltz, 1986; Thiriot-Quiévreux, 1986);
- assortative mating among adults (Zouros & Foltz, 1984a,b); and
- inbreeding.

Most of these explanations appear unlikely. Selection against heterozygotes produces an unstable polymorphism (Hedrick, 1983). Moreover, selection against heterozygotes in the larval stage can only be balanced against the apparent fitness advantages of heterozygotes in the juvenile and adult stages under rather restrictive theoretical conditions (Zouros & Foltz, 1984a,b). Geographic variation of allozyme frequencies is not large enough to account by the Wahlund Effect for the large deficiencies of heterozygotes observed (Koehn, 1975). Investigators have demonstrated null alleles for at least two allozymes in American oysters (Foltz, 1986), but the frequencies of null alleles do not appear to be high enough to account for the observed heterozygote deficiency (F.Sly & D.Hedgecock, unpublished data; Grady *et al.*, 1984). Assortative mating may not generate large enough heterozygote deficiencies, particularly if some matings are at random. Finally, dispersal of pelagic larvae ought to prevent co-settlement and eventual mating of close relatives. Thus, the explanation for heterozygote deficiencies in natural bivalve populations remains elusive, and the relationship of heterozygote deficiencies to maintenance of overall genetic diversity remains obscure. Curiously, heterozygote deficiency appears to coincide with heterozygosity-fitness correlation for reasons that are not yet clear (Gaffney *et al.*, 1990; Zouros, 1987).

HUMAN IMPOSED RISKS TO CONSERVATION OF GENETIC DIVERSITY

Habitat Destruction

Oyster habitats tend to be sites of dense human habitation and of activities such as fishing, aquaculture, recreation, and general shipping and commerce. Increasing human populations place ever increasing pressures on the quality of these habitats. Efforts to slow or reverse the most dramatic and visible trends have been successful recently. Bans in Britain and France and regulations in parts of the United States on the use of the antifouling compound tributyl tin (TBT) came about largely because of TBT's highly visible effects on benthic marine life, particularly on oyster populations (Alzieu, 1986; Champ & Lowenstein, 1987; Stebbing, 1985; Thain & Waldock, 1986).

Less dramatic or more slowly acting agents of environmental deterioration are less likely to arouse public outcry or government action (Clark, 1959). Production of oysters along the Atlantic and Gulf coasts of the United States has been declining for several decades, and in recent years east coast demand has been met in part by imports of Pacific oyster products from the west coast (Jackson, 1988). San Francisco Bay, which once produced over 2 million pounds of shucked oyster meat, now produces nothing (Barrett, 1963), and the accumulation of nonpoint source pollution in Puget Sound because of increased human populations in upland areas has led to the closure of progressively larger areas to shellfish culture (Leonard & Slaughter, 1990). Such changes are imperceptibly slow on a political or economic time scale, but are so instantaneous on an evolutionary time scale as to preclude genetic adaptations. Indeed, the irreversible loss of genetic resources through chance extinction can result simply from reducing a population below a demographically stable size (Goodman, 1987; Lande, 1988).

Overharvesting and Overplanting

Human predation is also responsible for declines in native oyster stocks. Reference has already been made to the decline of the flat oyster in Europe (Clark, 1959). The flat oyster native to western North America, *Ostrea lurida* or the Olympic oyster, was also severely devastated by gold rush immigrants in the mid- to late nineteenth century (Barrett, 1963). While ascribing the collapse of other bivalve fisheries solely to overharvesting may be difficult (Bourne, 1986), with the exception of the New Zealand fishery for the flat oyster *Tiostrea lutaria*, fisheries for wild oyster stocks today comprise only a minor fraction of world production (Bourne, 1986).

Coupled with periods of naturally poor recruitment or disease epidemics, overfishing can collapse the age structure of populations. This, in turn, can compromise the reproductive output needed to overcome high early mortality by removing the largest and most fecund broodstock from the population. Sex ratios might also be influenced by the protandric tendency of the hermaphroditic oyster.

Cultivation of wild or hatchery-produced seed promises to alleviate pressure on native stocks, but lack of regulation can lead to a tragedy of the commons, in which an area's overall health or productivity is sacrificed for short-term individual gains. So many oysters have been planted in the vicinity of the Isle de Oleron, France, for example, that natural phytoplankton supplies have been exceeded, oyster growth rates have slowed, and yield has declined (Heral *et al.*, 1986). In Japan, the growth and survival of oysters cultured in some bays have declined owing in part to human pollution, but also in part to excessive loading with oyster waste products (Ventilla, 1984).

Introductions of the Pacific Oyster

Compelling morphological, physiological, behavioural (Ranson, 1948, 1960, 1967; Yonge, 1960), and genetic evidence (Imai & Sakai, 1961;

Mathers *et al.*, 1974; Menzel, 1974) suggests that the Portuguese oyster, *Crassostrea angulata* is conspecific with *C. gigas*. Investigators have advanced three explanations for the existence of conspecific populations on opposite sides of the world:

- the Portuguese and Japanese oysters separated from the ancestral *C. gryphoides* in the Miocene and have remained unchanged (Dean, 1893; Korringa, 1952; Ranson, 1948, 1960; Stenzel, 1971);
- Portuguese traders may have introduced the Japanese oyster to Portugal (Ranson, 1948, 1960; Yonge, 1960) or vice versa;
- the Portuguese oyster was introduced to the western Pacific Ocean in the late 16th or early 17th centuries (Menzel, 1974).

The first of these hypotheses must be rejected because of the absence of morphological, reproductive, and molecular evolution between populations supposedly isolated for 10 million years. Menzel (1974) argued for the third alternative, being impressed by the coincidence that oyster farming in Japan dates back to the first half of the 17th century (Cahn, 1950; Korringa, 1976; Menzel, 1974; Ventilla, 1984). Shells resembling the Pacific oyster are present in prehistoric shell middens along the coasts of the Japanese archipelago, but these shells might represent one or more morphologically similar native species that were displaced by the introduced Portuguese oyster. There are several possible candidates (Hirase, 1932), including the Kumamoto variety of the Pacific oyster. The absence of a fossil record for *C. angulata* before European world explorations, however, makes the last hypothesis appear unlikely (Edwards, 1976; Ranson, 1948, 1960). Thus, the available evidence favours the second hypothesis, that Portuguese traders, either purposely or inadvertently by transport on the hulls of their ships, introduced *C. gigas* to Europe. Studies of DNA sequences may provide relevant new data.

The Pacific oyster has now been transplanted from Japan to all continents except Antarctica (Mann, 1979). The most important centre of production of Pacific oysters outside Japan is the northwest coast of North America, where Pacific oysters were introduced for commercial cultivation in the 1920s and 1930s (Barrett, 1963; Quayle, 1969). The combined worldwide production of *C. gigas* today far exceeds that of any other species of oyster (Bourne, 1986). How has transplantation of *C. gigas* for the purpose of cultivation affected the genetic diversity of this and other species of oyster?

In some cases, introduced Pacific oysters compete with native species of oysters, either on the basis of their own considerable abilities for colonization and ecological dominance, or because local aquaculturists, preferring their faster growth rates, encourage their spread. The Portuguese oyster supplanted the flat oyster along much of the west coast of France after introduction to the Gironde Estuary in the mid-1800s (Clark, 1959). These populations, which died out from disease in the early 1970s, were in turn replaced by imported Pacific oysters. Within a few years of introduction, Pacific oysters became established in Australia (Thomson, 1952, 1959), where they now threaten native *Saccostrea commercialist* populations (K.Chew, University of Washington at Seattle, pers. comm., September 1989). Since its introduction to New Zealand in 1971, probably *via* ship ballast, *C. gigas* has spread rapidly both by natural larval dispersal and by the deliberate planting activities of oyster growers, who prefer it to the native *Saccostrea glomerata* (Dinamani, 1971; Smith *et al.*, 1986). Extinction of native oysters with the attendant loss of their genetic resources is one possible consequence of transplantations.

The genetic consequences for Pacific oysters of introductions by man are apparently favourable. Genetic diversity of this species ought to be increased by the creation of new semi-isolated subpopulations, although diversity within subpopulations might be reduced if effective population sizes are restricted. Electrophoretic evidence indicates that cultivated populations of Pacific oysters in Japan are slightly more diverged from each other and from wild populations than are the wild populations from each other (Ozaki & Fujio, 1985). Genetic studies of Pacific oyster stocks introduced to the United Kingdom (Gosling, 1982) and to New Zealand (Smith *et al.*, 1986) reveal little loss of average heterozygosity, but losses of alleles and shifts in allozyme frequencies may reflect reduced effective population sizes in hatchery broodstock.

Impacts of Hatchery Propagation

Pacific oyster populations along the west coast of North America make an excellent case study of a marine species whose genetic resources are highly managed. The Pacific oyster is not managed as a fishery, but is primarily managed by the private commercial culture industry. Genetic resources are in the hands of a few large commercial hatcheries that dominate the seed oyster market. This resource management role of private business was consolidated following the rapid adoption by the culture industry in the 1980s of hatchery produced, 'eyed-larval' seed that can be set at remote farm sites (Chew, 1984; Jones & Jones, 1983). Prior to this, the industry obtained seed from a variety of sources, including shipments from Japan, collections from west coast bays where the species has naturalized to some extent, and hatcheries. The genetic status of naturalized populations relative to the now greater number of hatchery produced stocks is poorly understood.

Concern for the genetic resources of west coast Pacific oysters stems from a genetic study of natural seed set collected in Dabob Bay, and of two hatchery stocks derived from Dabob Bay natural stock and isolated from this wild progenitor stock and from each other for three generations (Hedgecock & Sly, 1990). Genetic differences among individuals were scored for 14 polymorphic enzymes. Overall levels of polymorphism and average heterozygosity were similar among the three population samples, but the hatchery populations had fewer alleles and had allelic frequencies that were substantially different from those in the wild stock (Table 6.3).

Assuming that allelic frequencies in the Dabob Bay sample were representative of those in the progenitors of the hatchery stocks and that the allozymes studied were selectively neutral, variances of allelic frequencies between wild progenitor and derived hatchery populations were used to estimate the temporal variances in allelic frequencies owing to random genetic drift (according to Pollak, 1983). Based on the inverse relationship between the magnitude of random genetic drift and effective population size, the per generation effective population sizes of the two hatchery stocks were calculated to be 40.6±13.9 (standard deviation) and 8.9±2.2 (standard deviation) (Table 6.4), far less than what would retard erosion of genetic diversity. These population size estimates account for the loss of alleles in the hatchery-propagated stocks, providing an independent test of the assumption that the allozymes were selectively neutral.

Applying these same methods to genetic data for other hatchery stocks of bivalves (e.g., Dillon & Manzi, 1987; Gosling, 1982; Vrijenhoek & Ford, 1987) yields similarly small estimates of effective population sizes (Hedgecock & Sly, 1990, and unpublished observations). This suggests that commercial bivalve hatcheries are ineffective in safeguarding genetic resources. Until hatchery management practices change, cultivated oyster populations, such as the Pacific oyster stocks along the west coast of North America, may be in danger of becoming inbred.

The status of genetic resources in hatchery-propagated stocks may also have implications for natural populations. Hatchery stocks reared on oyster beds are usually able to spawn in at least two successive summer seasons before being harvested. Thus, in bays where larval recruitment is possible, progeny of hatchery stocks may recruit to, and eventually swamp, native stocks. The contribution of larvae from hatchery-spawned oysters to 'natural set' is currently unknown. This scenario of hatchery stocks swamping native stocks is particularly troublesome given recent research by the State of Washington on enhancement of Puget Sound oyster beds with hatchery seed (D.S.Thompson, Washington Department of Fisheries, pers.comm., September 1989). One remedy for the situation, replenishment of genetic resources with fresh shipments of broodstock from Japan, may be restricted by regulations designed to prevent introduction of exotic pests and pathogens. Moreover, the spread of hatchery seed in Japan may be compromising the genetic diversity of those stocks as well (Ozaki & Fujio, 1985). Restocking would also undo any progress toward domestication that the west coast cultured population has made. Clearly, the better remedy would be for hatcheries to adopt broodstock management programmes that preserve genetic diversity.

The principal conservation need for west coast Pacific oysters, at this point, is sound information on the genetic structure of native and cultivated stocks. Allozyme frequency data for discrete stocks must be collected over time to measure genetic drift and permit estimation of effective population sizes. In addition, methods for nuclear and mitochondrial DNA sequence analysis that promise even greater

Table 6.3. Examples of genetic divergence among hatchery and wild Pacific oyster stocks. Allelic frequencies are given for 5 of 14 allozyme-coding loci studied (from Hedgecock & Sly, 1990).

Locus	Allele	Willapa (hatchery)	Dabob Bay (wild)	Humboldt (hatchery)
*AAT**	N	58	59	59
	108	0.172	0.398	0.000
	100	0.741	0.551	0.000
	92	0.086	0.051	1.000
*ACON -1**	N	58	47	59
	105	0.017	0.021	0.000
	103	0.181	0.213	0.000
	100	0.552	0.618	0.898
	97	0.250	0.085	0.102
*DIA**	N	57	51	59
	103	0.000	0.059	0.000
	100	0.772	0.578	0.161
	96	0.175	0.304	0.576
	94	0.053	0.049	0.263
	90	0.000	0.010	0.000
*TAP-2**	N	58	59	48
	102	0.009	0.153	0.026
	100	0.931	0.797	0.886
	98	0.043	0.051	0.088
	94	0.017	0.000	0.000
*TAP-3**	N	7	59	48
	102	0.053	0.119	0.052
	100	0.711	0.788	0.583
	98	0.184	0.076	0.333
	94	0.053	0.017	0.031

N is sample number.

resolution of population genetic parameters, such as variance in reproductive success, must be applied. The genetic impacts of commercial hatchery propagation upon future generations of Pacific oysters can and must be predicted.

Table 6.4. Estimates of effective population sizes for two commercial hatchery stocks of Pacific oysters. (From Hedgecock & Sly, 1990, calculated according to Pollak, 1983.)

	Willapa				Humboldt			
Locus	K	F_k	s.v.	N_k	K	F_k	s.v.	N_k
*AAT**	3	0.126	0.017	13.7	3	1.806	0.017	0.8
*ACON-1**	4	0.065	0.019	32.7	4	0.177	0.019	9.5
*ACON-2**	4	0.029	0.017	131.6	5	0.141	0.017	12.1
*APH**	2	0.047	0.017	49.5	2	0.068	0.017	29.4
*DIA**	5	0.066	0.019	31.7	5	0.268	0.018	6.0
*DAP**	5	0.031	0.017	109.7	5	0.072	0.020	28.4
*GPI**	6	0.042	0.017	60.0	7	0.094	0.017	19.5
*LAP**	5	0.043	0.017	57.9	5	0.046	0.017	52.0
*TAP-2**	4	0.093	0.017	19.8	3	0.105	0.017	17.2
*TAP-3**	4	0.062	0.017	33.7	4	0.148	0.019	11.6
*MDH-1**	2	0.156	0.017	10.8	2	0.034	0.017	88.0
*MDH-2**	2	0.000	0.017	-88.3	2	0.016	0.017	1580.4
*MPI**	2	0.012	0.017	-272.5	2	0.068	0.017	29.4
*SOD-2**	4	0.016	0.017	-1203.4	4	0.045	0.017	54.1
Totals/means:								
	51	0.054	40.6±	13.9	53	0.186	8.9±	2.2

Calculated for each allele *(K)* are:

variances (F^k) between each stock and the Dabob Bay WA wild stock from which they were derived; a correction for sampling variance (s.v.= $(1/2S_0 + 1/2S_1)$ where S_0 is number of individuals sampled from Dabob, and S_1 is number sampled from hatchery stock); and the estimated effective population sizes (N_k).

(F_k)s averaged over all alleles are given at thew bottom, together with estimates of (N_k) and standard deviations based on average (F_k)s and weighted *(by K)* mean s.v.s.

REFERENCES

Ahmed, M. 1975. Speciation in living oysters. *Advances in Marine Biology* **13**, 357-397.

Alzieu, C. 1986. TBT detrimental effects on oyster culture in France - evolution since anti-fouling paint regulation. In *Proceedings, Oceans 1986 Conference*, **4,** 1130-1134. Washington, D.C., September 23-25, 1986, Organotin Symposium.

Barrett, E.M. 1963. The California oyster industry. *California Department of Fish and Game Fish Bulletin* **123**, 1-103.

Beattie, J.H., McMillin, D. & Weigardt, L. 1982. The Washington state oyster industry: A brief overview. In *Proceedings of the North American Oyster Workshop* (Chew, K.K. ed.) *World Mariculture Society, Special Publication* **1**, 28-38.

Beattie, J.H., Davis, J.P., Downing, S.L. & Chew, K.K. 1988. Summer mortality of Pacific oysters. In *Disease Processes in Marine Bivalve Molluscs* (Fisher, W.S. ed.) *American Fisheries Society Special Publication* **18**, 265-268.

Blanc, F. & Bonhomme, F. 1987. Polymorphisme génétique des populations naturelles de mollusques d'interêt aquicole. In *Selection, Hybridization and Genetic Engineering in Aquaculture* (Tiews, K. ed.) Vol.**I**, 59-78. Berlin: Heenemann.

Bonar, D.B., Coon, S.L.,Walch, M.,Weiner, R.M. & Fitt, W. 1990. Control of oyster settlement and metamorphosis by endogenous and exogenous chemical cues. *Bulletin of Marine Science* **46**, 484-498.

Bourne, N. 1986. Bivalve fisheries: Their exploitation and management with particular reference to the northeast Pacific region. In *North Pacific Workshop on Stock Assessment and Management of Invertebrates* (Jamieson, G.S. & Bourne, N. eds.) *Canadian Special Publication in Fisheries & Aquatic Sciences* **92**, 2-13.

Buroker, N.E. 1983. Population genetics of the American oyster *Crassostrea virginica* along the Atlantic coast and the Gulf of Mexico. *Marine Biology* **75**, 99-112.

Buroker, N.E. 1984. Gene flow in mainland and insular populations of *Crassostrea* (Mollusca). *Biological Bulletin* **166**, 550-557.

Buroker, N.E., Hershberger, W.K. & Chew, K.K. 1979. Population genetics of the family Ostreidae. I. Intraspecific studies of *Crassostrea gigas* and *Saccostrea commercialist*. *Marine Biology* **54**, 157-169.

Cahn, A.R. 1950. Oyster culture in Japan. *U.S. Fish & Widlife Service Bureau,* Commercial Fisheries Fish Leaflet **383**, 1-80.

Champ, M.A. & Lowenstein, F.L. 1987. TBT: The dilemma of high-technology antifouling paints. *Oceanus* **30**, 69-77.

Clark, E. 1959. *The Oysters of Locmariaquer.* New York: Pantheon Books.

Coe, W.R. 1934. Alternation of sexuality in oysters. *American Naturalist* **68**, 236-251.

Coon, S.L., Bonar, D.B. & Weiner, R.M. 1985. Induction of settlement and metamorphosis of the Pacific oyster, *Crassostrea gigas* (Thunberg), by L-dopa and catecholamines. *Journal of Experimental Marine Biology & Ecology* **94**, 211-221.

Chew, K.K. 1984. Recent advances in the cultivation of molluscs in the Pacific United States and Canada. *Aquaculture* **39**, 69-81.

Dayton, P.K. 1979. Ecology: Science or religion? In *Ecological Processes in Coastal and Estuarine Systems* (Livingston, R. ed.) pp.3-17. New York: Plenum Press.

Dean, B. 1893. Report on European methods of oyster culture. Bulletin of the U.S. Fish Commission **11**, 357-406.

Dillon, R.T. & Manzi, J.J. 1987. Hard Clam, *Mercenaria mercenaria*, broodstocks: Genetic drift and loss of rare alleles without reduction in heterozygosity. *Aquaculture* **60**, 99-105.

Dinamani, P. 1971. Occurrence of the Japanese oyster, *Crassostrea gigas* (Thunberg), in Northland, New Zealand. *New Zealand Journal of Marine & Freshwater Research* **5**, 352-357.

Edwards, C. 1976. A study in erratic distribution: The occurrence of the medusa *Gonionemus* in relation to the distribution of oysters. *Advances in Marine Biology* **14**, 251-284.

Foltz, D.W. 1986. Null alleles as a possible cause of heterozygote deficiencies in the oyster *Crassostrea virginica* and other bivalves. *Evolution* **40**, 869-870.Fujio, Y. 1979. Enzyme polymorphism and population structure of the Pacific oyster *Crassostrea gigas*. *Tohuku Journal of Agricultural Research* **30**, 32-42.

Gaffrey, P.M., Scott, T.M., Koehn, R.K. & Diehl, W.J. 1990. Interrelationships of heterozygosity, growth rate, and heterozygote deficiencies in the coot clam, *Mulinia lateralis*. *Genetics* **124**, 687-699.

Gaines, S.D. & Roughgarden, J. 1985. Larval settlement rate: A leading determinant of structure in an ecological community of the marine intertidal zone. *Proceedings of the National Academy of Science of the USA* **82**, 3707-3711.

Gaines, S.D., Brown, S. & Roughgarden, J. 1985. Spatial variation in larval concentrations as a cause of spatial variation in settlement for the barnacle, *Balanus glandula*. *Oecologia* **67**, 267-272.

Galtsoff, P.S. 1961. Physiology of reproduction in molluscs. *American Zoologist* **1**, 273-289.

Galtsoff, P.S. 1964. The American Oyster *Crassostrea virginica* Gmelin. *Fishery Bulletin* **64**, 1-480.

Gjedrem, T. 1983. Genetic variation in quantitative traits and selective breeding in fish and shellfish. *Aquaculture* **33**, 51-72.

Goodman, D. 1987. The demography of chance extinction. In *Viable Populations for Conservation* (Soulé, M.E. ed.) pp.11-34. Cambridge: Cambridge University Press.

Gosling, E.M. 1982. Genetic variability in hatchery-produced Pacific oyster (*Crassostrea gigas* Thunberg). *Aquaculture* **26**, 273-287.

Grady, J.M., Soniat, T.M. & Rogers, J.S. 1984. Genetic variability and gene flow in populations of *Crassostrea virginica* (Gmelin) from the northern Gulf of Mexico. *Journal of Shellfish Research* **8,** 227-232.

Grizel, H., Mialhe, E., Chagot, D., Buolo, V. & Bachere, E. 1988. Bonamiasis: A model study of diseases in marine molluscs. In *Disease Processes in Marine Bivalve Molluscs* (Fisher, W.S. ed.) *American Fisheries Society Special Publication* **18**, 1-4.

Groue, K.J. & Lester, L.J. 1982. A morphological and genetic analysis of geographic variation among oysters in the Gulf of Mexico. *Veliger* **24**, 331-335.

Haley, L.E. 1977. Sex determination in the American oyster. *Journal of Heredity* **68**, 114-116.

Haley, L.E. 1979. Genetics of sex determination in the American oyster. *Proceedings of the National Shellfish Association* **69**, 54-57.

Hedgecock, D. 1982. Genetical consequences of larval retention: Theoretical and methodological aspects. In *Estuarine Comparisons* (Kennedy, V.S. ed.) pp. 553-568. New York: Academic Press.

Hedgecock, D. 1986. Is gene flow from pelagic larval dispersal important in the adaptation and evolution of marine invertebrates? *Bulletin of Marine Science* **39**, 550-564.

Hedgecock, D. 1988. Can the Pacific oyster be domesticated? In *West Coast Mollusc Culture: A Present and Future Perspective* (Amadei, R. ed.). Report No. **T-CSGCP-017**, 69-72. La Jolla: California Sea Grant College Program.

Hedgecock, D. & Berthelemy-Okazaki, N. 1984. Genetic diversity within and between populations of American oysters (*Crassostrea*). *Malacologia* **25**, 535-549.

Hedgecock, D. & Sly, F.L. 1990. Genetic drift and effective population sizes of hatchery-propagated stocks of the Pacific oyster *Crassostrea gigas*. *Aquaculture* **88**, 21-38.

Hedgecock, D., Tracey, M.L. & Nelson, K. 1982. Genetics. In *The Biology of Crustacea* (Abele, L.G. ed.) **2**, 297-403. New York: Academic Press.

Hedrick, P.W. 1983. *Genetics of Populations*. New York: Van Nostrand Reinhold.

Héral, M., Deslous-Paoli, J.-M. & Prou, J. 1986. Dynamiques des productions et des biomasses des huîtres creuses cultivées (*Crassostrea angulata* et *crassostrea gigas*) dans le bassin de marenne-oléron depuis un siècle. *International Council for the Exploration of the Sea* **CM.1986/F:41**.

Hershberger, W.K., Perdue, J.A. & Beattie, J.H. 1984. Genetic selection and systematic breeding in Pacific oyster culture. *Aquaculture* **39**, 237-245.

Hillman, R.E. 1964. Chromatographic evidence of intraspecific genetic differences in the eastern oyster, *Crassostrea virginica*. Systematic Zoology **13**, 12-18.

Hillman, R.E. 1965. Chromatographic studies of allopatric populations of the Eastern oyster, *Crassostrea virginica*. *Chesapeake Science* **6**, 115-116.

Hirase, S. 1932. Some species of Japanese oysters. *Japanese Journal of Zoology* **4**, 213-222.

Imai, T. & Sakai, S. 1961. Study of breeding of Japanese oyster, *Crassostrea gigas. Tohoku Journal of Agricultural Research* **12**, 125-171.

Jackson, D.D. 1988. Is the oyster on its last legs? *Smithsonian* **18**, 61-70.

Jones, G. & Jones, B. 1983. Methods for setting hatchery produced oyster larvae. *British Columbia Ministry of Environment, Marine Resources Branch, Information Report* **4.**

Koehn, R.K. 1975. Migration and population structure in the pelagically dispersing marine invertebrate, *Mytilus edulis*. In *Isozymes IV: Genetics and Evolution* (Markert, C.L. ed.) pp.945-959. New York: Academic Press.

Koehn, R.K. 1984. Adaptive aspects of biochemical and physiological variability. In Proceedings of the 19th European Marine Biology Symposium (Gibbs, P.E. ed.) pp. 945-959. Cambridge: Cambridge University Press.

Koehn, R.K. 1990. Heterozygosity and growth in marine bivalves: Comments on the paper by Zouros, Romero-Dorey, and Mallet (1988). *Evolution* **44**, 213-216.

Korringa, P. 1952. Recent advances in oyster biology. *Quarterly Review of Biology* **27**, 266-308, 339-365.

Korringa, P. 1976. *Farming the Cupped Oysters of the Genus Crassostrea.* Amsterdam: Elsevier.

Lande, R. 1988. Genetics and demography in biological conservation. Science **241**, 1455-1460.

Lande, R. & Barrowclough, G.F. 1987. Effective population size, genetic variation, and their use in population management. In *Viable Populations for Conservation* (Soulé, M.E. ed.) pp.87-123. Cambridge: Cambridge University Press.

Lannan, J.E. 1980a. Broodstock management of *Crassostrea gigas*: I. Genetic variation in survival in the larval rearing system. *Aquaculture* **21**, 323-336.

Lannan, J.E. 1980b. Broodstock management of *Crassostrea gigas*: IV. Inbreeding and larval survival. *Aquaculture* **21**, 352-356.

Lannan, J.E., Robinson, A. & Breese, W.P. 1980. Broodstock management of *Crassostrea gigas*: II. Broodstock conditioning to maximize larval survival. *Aquaculture* **21**, 337-345.

Leonard, D.L. & Slaughter, E.A. 1990. *The quality of shellfish growing waters on the west coast of the United States.* Rockville, Md.: National Oceanographic and Atmosphere Administration.

Li, M.F., Fleming, C. & Stewart, J.E. 1967. Serological differences between two populations of oysters (*Crassostrea virginica*) from the Atlantic coast of Canada. *Journal of the Fisheries Research Board of Canada* **24**, 443-446.

Longwell, A.C. & Stiles, S.S. 1973. Gamete cross incompatibility and inbreeding in the commercial American oyster *Crassostrea virginica* Gmelin. *Cytologia* **38**, 521-533.

Loosanoff, V.L. & Nomejko, C.A. 1951. Existence of physiologically-different races of oysters, *Crassostrea virginica*. *Biological Bulletin* **101**, 151-156.

Mallet, A.L. & Haley, L.E. 1983. Effects of inbreeding on larval and spat performance in the American oyster. *Aquaculture* **33**, 229-235.

Mallet, A.L., Zouros, E., Gartner-Kepkay, K.E., Freeman, K.R. & Dickie, L.M. 1985. Larval viability and heterozygote deficiency in populations of marine bivalves: Evidence from pair matings in mussels. *Marine Biology* **87**, 165-172.

Mann, R. 1979. *Exotic Species in Mariculture*. Cambridge: The MIT Press.

Mathers, N.F., Wilkins, N.P. & Walne, P.R. 1974. Phosphoglucose isomerase and esterase phenotypes in *Crassostrea angulata* and *C. gigas*. *Biochemical Systematics & Ecology* **2**, 93-96.

Menzel, R.W. 1951. Early sexual development and growth of the American oyster in Louisiana waters. *Science* **113**, 719-721.

Menzel, R.W. 1956. The effect of temperature on the ciliary action and other activities of oysters. *Florida State University Studies* **22**, 25-36.

Menzel, R.W. 1968. Chromosome numbers in nine families of marine pelecypod mollusks. *Nautilus* **82**, 45-58.

Menzel, R.W. 1974. Portuguese and Japanese oysters are the same species. *Journal of the Fisheries Research Board of Canada* **31**, 453-456.

Mitton, J.B. & Grant, M.C. 1984. Associations among protein heterozygosity, growth rate, and developmental homeostasis. *Annual Review of Ecology & Systematics* **15**, 479-499.

Nevo, E., Beiles, A. & Ben-Shlomo, R. 1984. The evolutionary significance of genetic diversity: Ecological, demographic and life history correlates. In *Evolutionary Dynamics of Genetic Diversity* (Mani, G.S. ed.) *Lecture Notes in Biomathematics* **53**, 13-213 Berlin: Springer-Verlag.

Numachi, K. 1962. Serological studies of species and races of oysters. American Naturalist **96**, 211-217.

Numachi, K. 1971. Japanese oyster species, breed and distribution. In *Aquaculture in Shallow Seas: Progress in Shallow Sea Culture* (Imai, T. ed.) p.88. Tokyo: Koseisya-Koseikakau (in Japanese).

Ozaki, H. & Fujio, Y. 1985. Genetic differentiation in geographical populations of the Pacific oyster (*Crassostrea gigas*) around Japan. *Tohuku Journal of Agricultural Research* **36**, 49-61.

Palumbi, S.R. & Wilson, A.C. 1990. Mitochondrial DNA diversity in the sea urchins *Stronglyocentrotus purpuratus* and *S. droebachiensis*. *Evolution* **44**, 403-415.

Pollak, E. 1983. A new method for estimating the effective population size from allele frequency changes. *Genetics* **104**, 531-548.

Quayle, D.B. 1969. Pacific oyster culture in British Columbia. *Fisheries Research Board of Canada Bulletin* **169**.

Ranson, G. 1948. Prodissoconques et classification des Ostréides vivants. *Bulletin Musée Royale d'Histoire Naturelle Belgique* **24**, 1-12.Ranson, G. 1960. Les prodissoconques (coquilles larvaires) des Ostréides vivants. *Bulletin de l'Institut Océanographique (Monaco)* **56**, 1-41.

Ranson, G. 1967. Les especes d'huîtres vivant actuellement dans le monde, définies par leurs coquilles ou prodissoconques. *Revue des Travaux de l'Institut des Pêches Maritimes, Paris* **31**, 1-146.

Reeb, C.A. & Avise, J.C. 1990. A genetic discontinuity in a continuously distributed species: Mitochondrial DNA in the American oyster, *Crassostrea virginica*. *Genetics* **124**, 397-406.

Rose, R.L. 1984. Genetic variation in the oyster, *Crassostrea virginica* (Gmelin), in relation to environmental variation. *Estuaries* **7**, 128-132.

Roughgarden, J., Gaines, S. & Possingham, H. 1988. Recruitment dynamics in complex life cycles. *Science* **241**, 1460-1466.

Schaal, B.A. & Anderson, W.W. 1974. An outline of techniques for starch gel electrophoresis of enzymes from the American oyster *Crassostrea virginica* Gmelin. *Technical Report Series of the Georgia Marine Science Center* **74(3)**, 1-17.

Singh, S.M. & Green, R.H. 1984. Excess of allozyme homozygosity in marine molluscs and its possible biological significance. *Malacologia* **25**, 569-581.

Smith, P.J., Ozaki, H. & Fujio, Y. 1986. No evidence for reduced genetic variation in the accidentally introduced oyster *Crassostrea gigas* in New Zealand. *New Zealand Journal of Marine & Freshwater Research* **20**, 569-574.

Stauber, L.A. 1947. On possible physiological species in the oyster, *Ostrea virginica*. *Anatomical Record* **94**, 614.

Stauber, L.A. 1950. The problem of physiological species with special reference to oysters and oyster drills. *Ecology* **31**, 109-118.

Stebbing, A.R.D. 1985. Organotins and water quality - some lessons to be learned. *Marine Pollution Bulletin* **16**, 383-390.

Stenzel, H.B. 1971. Oysters. In *Treatise on Invertebrate Paleontology* (Moore, R.C. ed.) Part **N**, 953-1224. Boulder, Colo.: Geological Society of America.

Thain, J.E. & Waldock, M.J. 1986. The impact of tributyl tin (TBT) antifouling paints on molluscan fisheries. *Water Science & Technology* **18**, 193-202.

Thiriot-Quiévreux, C. 1986. Etude de l'aneuplöide dans différents naissains d'Ostreidae (Bivalvia). *Genetica* **70**, 225-231.

Thomson, J.M. 1952. The acclimatization and growth of the Pacific oyster (*Gryphaea gigas*) in Australia. *Australian Journal of Marine & Freshwater Research* **3**, 64-73.

Thomson, J.M. 1959. The naturalization of the Pacific oyster in Australia. *Australian Journal of Marine & Freshwater Research* **10**, 144-149.

Underwood, A.J. & Denley, E.J. 1984. Paradigms, explanations, and generalizations in models for the structure of intertidal communities on rocky shores. In *Ecological*

Communities (Strong, D.R., Simberloff, D., Abele, L. & Thistle, A. eds.) pp.151-180. Princeton, N.J.: Princeton University Press.

Ventilla, R.F. 1984. Recent developments in the Japanese oyster culture industry. *Advances in Marine Biology* **21**, 1-57.

Vrijenhoek, R. & Ford, S. 1987. Maintenance of heterozygosity in oysters during selective breeding for tolerance to MSX infections. *Journal of Shellfish Research* **7**, 179.

Wright, S. 1931. Evolution in Mendelian populations. *Genetics* **16**, 97-159.

Wright, S. 1969. *Evolution and the Genetics of Populations, Vol. II. The Theory of Gene Frequencies.* Chicago: Chicago University Press.

Yonge, C.M. 1960. *Oysters.* London: Collins.

Zouros, E. 1987. On the relation between heterozygosity and heterosis: An evaluation of the evidence from marine mollusks. In *Isozymes: Current Topics in Biological and Medical Research* (Rattazi, M.C., Scandalios, J.G. & Whitt, G.S. eds.) **13**, 255-270 New York: Liss.

Zouros, E. 1990. Heterozygosity and growth in marine bivalves: Response to Koehn's remarks. *Evolution* **44**, 216-218.

Zouros, E. & Foltz, D.W. 1984a. Possible explanations of heterozygote deficiency in bivalve molluscs. *Malacologia* **25**, 583-591.

Zouros, E. & Foltz, D.W. 1984b. Minimal selection requirements for the correlation between heterozygosity and growth, and for the deficiency of heterozygotes, in oyster populations. *Developmental Genetics* **4**, 393-405.

Zouros, E., & Foltz, D.W. 1987. The use of allelic isozyme variation for the study of heterosis. In *Isozymes: Current Topics in Biological and Medical Research* (Rattazi, M.C., Scandalios, J.G. & Whitt, G.S. eds.) **13**, 1-59. New York: Liss.

7
THE COMMON CARP AND CHINESE CARPS

Giora W.Wohlfarth

The author of *Paradise Lost* ...attempted to justify God's ways to Man. He would have done a more useful Work in undertaking to explain the ways of God to Fish. Philosophers have wasted their own and their readers time in speculating upon the Immortality of the Soul; the Alchemists have pored for centuries over their crucibles in the vain hope of discovering the Elixir or the Stone. Meanwhile in every pond and river, one may find Carps that have outlived three Platos and half a dozen Paracelsuses. The Secret of eternal Life is not to be found in old Books, bor in liquid Gold, nor even in Heaven; it is to be found in the mud and only awaits a skilful Angler.

Aldous Huxley: *After Many a Summer*

INTRODUCTION

Fishes described in this chapter include the common carp (*Cyprinus carpio*), and a group of five cyprinids, collectively termed Chinese carps, namely: silver carp (*Hypophthalmichthys molitrix*), bighead carp (*Aristichthys nobilis*), grass carp (*Ctenopharyngodon idella*), black carp (*Mylopharyngodon piceus*), and mud carp (*Cirrhina molitorella*). The amount of information on common carp far exceeds that available for Chinese carps, probably because of its cosmopolitan distribution. Aquaculture in both China and Europe began as a monoculture of common carp. This may be as a result of its ease of spawning in captivity and of live transport over long distances. The total catch of these carps in 1990 was c. 4.5 million metric tons, a far greater total than any other group of freshwater fishes. Most of this catch came from Chinese aquaculture yields (Table 7.1). The existence of true natural stocks of these fishes is in doubt. The common carp has been transferred to most regions amenable to its propagation and growth, and become genetically adapted to these altered conditions. In many areas of the carp's endemic occurrence, aquaculture has been practised for generations, and wild stocks are suspect of being crossbred derivatives with domesticated migrantsfrom aquaculture. The large differences demonstrated between

Table 7.1. Nominal catches (x 1000 tons) of common carp and Chinese carps exceeding 10,000 tons per year in 1990. Aquaculture yields given in parentheses, when differing from nominal catches.

Country	Common carp		Silver carp	Bighead carp	Grass carp	Mud carp	Black carp
China	522.4		1398.9	658.2	1023.2	80	37.5
USSR	319.8	(300.0)	(90.1)				
Indonesia	101.6	(89.8)					
Egypt	31.0						
Iran	30.0	(15.2)	(15.5)				
Mexico	27.7	(1.4)					
Hungary	23.4	(13.2)					
Japan	22.6	(16.2)					
Poland	22.0						
Germany	20.1	(19.1)					
CSSR	18.8						
Bulgaria	18.8	(8.2)					
Turkey	17.0	(1.0)					
Romania	13.4	(**)					
Yugoslavia	13.2	(10.0)					
S. Korea	*	(10.8)					
Total:	1229.4	(1112.7)	1423.4 (1515.3) ***	677.7 (671.5)	1042.4 (1047.1) ***	80.3	37.5 (37.9) ***

* Nominal catch not available
** Aquaculture yield not available
*** Aquaculture yield higher than nominal catch data apparently due to mistake in FAO statistics.

domesticated carp and progenies of wild carp taken from the Danube and Lake Valence in Hungary, are not proof of stock purity of the wild carp. Common carp in North America, where they were introduced during the last century, have the typical appearance of wild carp, in both body shape and scale pattern. The original individuals introduced were taken from domesticated stocks, and the appearance of these 'wild' American carp is an example of their adaptation to natural conditions.

The genetic purity of wild stocks of Chinese carps is also in some doubt, in spite of their short history of domestication. Catches of fish from the wild are decreasing, apparently as a result of overfishing and destruction of spawning grounds. This is particularly evident for catches of fry of silver, bighead and grass carp (Li, 1989). Aquaculture yields of Chinese carps have been increasing in the last decade and this may lead to some genetic flooding through escape of fish from aquaculture facilities.

CLASSIFICATION

Common carp is the only generally recognized species of *Cyprinus*. Eleven further species or subspecies of this genus were described from Yunan, China, according to their skeletal characteristics (Wei & Xinlou, 1986). Division of the genus into *Mesocyprinus* and *Cyprinus* has also been suggested. Alikhuni (1966) enumerated 18 more *Cyprinus* species, also in the Far East, but Sarig (1966) regarded all differently named species of the genus as synonyms of *C. carpio*.

C. carpio has been divided into four subspecies differing in their distribution and number of chromosome arms (Rab *et al.*, 1989). A subclassification according to scale pattern into scaled carp, mirror carp and leather carp (as suggested by Alikhuni, 1966) is not justified, since all scale patterns may segregate from the same spawn, as a result of their simple Mendelian inheritance (see below). Here, the common carp is regarded as the only species in its genus, and without any subspecific divisions (Balon, 1974; Komen, 1990).

ORIGIN AND MIGRATION OF COMMON CARP

The common carp probably originated in Asia Minor and the area of the Caspian Sea at the end of the Pleistocene era (Steffens, 1966; Balon, 1974). From here it appears to have spread east and west during the last post-glacial period. In an easterly direction it reached China, but did not cross the Bering land bridge to America. Its westerly migration reached the Black and Aral Seas, arriving at the Danube River system some 10,000 years ago. The species did not extend north to Scandinavia and Britain (Balon, 1974).

There is a suspicion that carp were transported from the Danube area to Italy via the Amber Road by the Romans, and stocked into holding ponds (*piscinae*) (Balon, 1974; Berka, 1985), but this is doubted owing to their not being mentioned in classical Roman cookery books (Komen, 1990). The widespread distribution of common carp in Europe took place between the 7th and 13th centuries, originally to holding ponds in monasteries, perhaps because live carp can be transported easily. Here they served as an important food item on days of fasting, including over 100 days per year, when consumption of other meats was prohibited by religious regulations. Presumably fish farming in Europe evolved from these monastic holding ponds, after it was observed that carp could breed and grow in these ponds.

DOMESTICATION OF COMMON CARP

In Europe common carp were domesticated as a result of their adaptation to captive management conditions and the intentional selection of broodstock. The centre of carp cultivation in Europe was Bohemia and culture methods were described by Bishop Jan Dubrovius in his book *On Ponds*, published in the first half of the 16th century (Berka, 1985; Kozakava, 1987). Pond management was surprisingly advanced, including separate ponds for spawning and early rearing; for rearing two-year-old fish; and for growing fish to market size during a further 2-4 years. This resulted in a high value of ponds, a 260 hectare pond in the 16th century having a price

> equal to the value of two to three villages, with all serfs, land and inventory (Berka, 1985).

Similarly in Britain, knowledge of pond culture, mainly of common carp, was of a surprisingly high standard and it was a very profitable activity (Taverner, 1600; North, 1835). The influence of stocking rates, polyculture, manuring, and feeding were well understood. Domestication of the European carp was largely a result of captive adaptation to these conditions. Broodstock selection was carried out regularly in Europe, as will be described below.

Aquaculture has been practised widely in China for about 4000 years, much longer than in Europe. The first tract on the culture of fish was written by Faan Li in the 5th century BC. Originally it consisted of monoculture of the common carp. During the Tang Period, common carp became a symbol of the royal family, and when caught had to be released, under threat of punishment by caning (Committee for the Collection of Experiences in the Culture of Freshwater Fish Species in China, 1981). This led to the culture of Chinese carps, though at a later stage the common carp was re-introduced into this polyculture system. Domestication of common carp in China is a result of their adaptation to the captive conditions of Chinese aquaculture, which were (and still are) very different from those in Europe. Management practices included high density polyculture of Chinese and common carp and frequent intermittent harvesting by seining, the largest individuals being taken for marketing or consumption (Hoffmann, 1934). Stocking densities were uncontrolled, since common carp spawned in these production ponds. Nutrient inputs consisted largely of agricultural, domestic and industrial by-products (the term wastes is misplaced in the connotation of Chinese agriculture), such as manures, sewage, kitchen refuse and the remnants of silk-worm pupae. Supplemental feed consisted largely of grasses and snails gathered to feed grass carp and black carp directly and, since these carp are inefficient food utilisers, also other fishes in the pond indirectly. Domestication of common carp to these conditions was complicated by regular introductions of wild carp into aquaculture facilities. This involved augmenting breeding schools with fish caught in the wild (Hoffmann, 1934). Another Chinese practice consisted of placing spawning beds into suitable places in streams, and transferring them to ponds for hatching after spawning had taken place (Liu, 1941). Genetic improvement of common carp was not practised in traditional Chinese aquaculture (S.Y.Lin, pers. comm.), but some projects have been initiated recently. These involved comparisons among isolates collected in different rivers, and testing F1 crossbreds (Minghua *et al.*, 1992).

ORIGIN AND DOMESTICATION OF CHINESE CARPS

The earliest fossils of common and grass carp from the Pliocene era, and of the bighead and black carp from the Pleistocene, were found in Shansi Province (Committee for the Collection..., 1981). The origin of these fishes is in the major Chinese rivers, the Amur (Heilongjiang), Yangtse (Changjiang), and Pearl (Zhujiang) rivers, though there are no bighead and mud carp in the Amur (Li, 1989), the most northerly and coldest of these systems. Domestication of the Chinese carps is of relatively recent origin, since their commercial propagation began only in the middle of the present century (Tapiador *et al.*, 1977). Previous to that, ponds were stocked with fry captured in the wild (Hoffmann, 1934; Committee for the Collection...., 1981). The number of generations of reproduction in captivity cannot be large, considering the long generation interval of these fishes (3-7 years, according to species and sex under the most favourable environments). Surprisingly, the influence of this short domestication process seems to be negative, since progenies of silver and bighead carp from wild progenitors grew 5-10% faster than did those from hatchery parents (Li *et al.*, 1987a,b; Li, 1989).

NOMINAL CATCHES

The total catch of common and Chinese carps in 1990 was c. 4.5 million metric tons, over 80% coming from China (Table 7.1). The Chinese catch of Chinese carps represented about 98% of the world catch of these species. Common carp is the only one of these cyprinids with large catches outside China, with the exception of silver carp catches from the former USSR and Iran. It is also the only one in which there is any meaningful difference between nominal catch and aquaculture yield. Large non-aquaculture catches of common carp, presumably from natural waters were obtained in Indonesia, Japan, Hungary, Bulgaria, Turkey, Iran, and Mexico. In seven European countries, carp catch varied from 13,400-22,000 metric tons, and in some of these this represented well over half the total fish catch from inland waters.

INTRODUCTIONS OF THE COMMON CARP

The introduction of common and Chinese carps has been summarized by Welcomme (1988). The common carp has been introduced very widely (Table 7.2), in most cases for aquaculture, but sometimes for sport, supplying pituitary glands, or as ornamental fish. Spontaneous reproduction took place in most cases, but artificial reproduction was carried out in several countries, particularly in Latin America. Most of the introductions to Africa and Europe came from European countries, although the introduction to Egypt was from Indonesia. Introductions to Asian countries were mostly from Asia, and most of those to Latin America were from North America. The development of carp cultivation following these introductions has not always been well documented. Some examples follow.

Central Europe

Bohemia, which is in the area of common carp's natural distribution, is regarded as the historical centre of fish culture in Europe (Müller, 1989). The area of carp ponds increased to a maximum of 180,000 hectares by the 16th century (Berka, 1985; Zobel, 1989) but declined during the 17th century as a result of the ravages of the 30 Years War. In the following period aquaculture was considered ineffective for food production, and this attitude changed only in the second half of the 19th century.

Siberia

Carp culture in Siberia is difficult because of the low temperatures and the short growing season (Kirpichnikov, 1957; Anon, 1983b; Babushkin, 1987). Adaptation of the common carp to Siberian conditions by a specific breeding scheme resulted in the production of the Ropsha carp (see below).

Britain

Common carp was apparently introduced into Britain in the 15th century (O'Grady & Spillet, 1985) and used for producing table fish. In the old English literature (Taverner, 1600; North, 1835) fish cultivation in ponds (particularly that of common carp) was considered a highly profitable type of agriculture:

Table 7.2. Introductions of common carp (from Welcomme, 1988).

To	From	Year	Reason	Reproductio
Africa				
Cameroons	Israel	1970	A	+
C. African Rep.	Israel	1966	A	a
Egypt	Indonesia	1934	A	+
Ethiopia	?	1936	A	+
Ghana	Uganda	1962	A	+
Ivory Coast	Italy	1976	A	+
Kenya	Uganda	1969	A	+
Madagascar	?	1914	A	+
Mauritius	India	1976	A	+
Morocco	France	1925	A	+
Nigeria	Austria/Israel	1954, 1976	A	?
Rwanda	Israel	1960	A	+
South Africa	Germany	1859	?	?
Sudan	India	1975	P	?
Swaziland	South Africa	?	A	?
Togo	?	?	A	?
Uganda	Israel	1963	A	+
Zimbabwe	South Africa	1925	A	+
America				
Argentina	Brazil	?	A	+
Bolivia	Mexico	1945	A, O	+
Brazil	USA	1898, 1977	A	+
Canada	France	1831	?	+
Chile	Germany	1875	F	+
Colombia	?	1940	A	+
Costa Rica	China/Taiwan	1976	P	+
Cuba	USA/USSR	1927, 1983	A, sport	+
Dominican Republic	Mexico	1953	A	+
Ecuador	?	?	?	a
El Salvador	Guatemala	1965	A	+
Guatemala	?	1954	A	+
Haiti	Israel	?	?	a
Honduras	Nicaragua	1956	A	a
Mexico	France	1872/3	A	+
Nicaragua	Mexico	1964	A	a
USA/Israel/Colombia	Panama	1976, 1981	A	+
Peru	Japan/China	1946, 1960	A	+
Surinam	Japan	1968	A	a

Uruguay	Brazil	1850	straying +
USA	Germany	1877	food +
Venezuela	?	1940	F +
Asia			
Afghanistan	China	1970s	A +
India	Sri Lanka/Thailand	1939-1958	A +
Indonesia	China/Japan	1970s	A +
Korea	Israel	1973	A +
Malaysia	China	1980s	A +
Nepal	Hungary	?	A in ponds
Pakistan	Thailand	1964	A +
Philippines	China/Hong Kong	1910, 1915	A +
Sri Lanka	Europe	1913	F +
Thailand	China/Japan/Israel/ Germany	1913 - onwards	A +
	Europe		
Cyprus	Israel	1966	A +
Finland	Germany/Sweden/ USSR	1955, 1959 & 1961	A in ponds
Greece	?	?	new stock +
Israel	Europe	1931	A +
Sweden	?	?	A +
Oceania			
Australia	UK	1900s	? +
Fiji	New Zealand	1936	O ?
Guam	?	?	? +
Hawaii	?	1900s	? a
New Zealand	UK	1864-1911	? +
Papua New Guinea	Australia	1959	A +

A: aquaculture; a: artificially; F: fisheries; O: ornamental; P: pituitaries.

> I do not think ground would yeeld unto the owner any other way so much benefite as to be converted into such ponds (Taverner, 1600).

Present day carp production is on a very small scale, mainly for

> restocking, recreational fisheries and to supply the increasing ethnic market (O'Grady & Spillet, 1985).

France

Carp is the most common fish in French warm water aquaculture. Spawning is induced artificially in warm water facilities during May. The fish are grown mostly to weights of 25-250 grammes for stocking as a sport fish. Some are grown to >2 kilogrammes as table fish for export. Only a few carp are eaten in France (Huner, 1985).

USA and Canada

Common carp were introduced by the US Fish Commission in 1877 for distribution to potential growers, though they had been grown previously in New York since the 1830s

(Cooper, 1987). Carp in Canada originated from US introductions, either through intentional introductions or by invasion (MacCrimmon, 1968). Initially these introductions appeared to be very successful. The fish reproduced and spread and were utilized as food on a large scale.

> Carp have found their way into every city and town of the west and on almost every table. (Bartlett, 1905)

Capture, processing, transport, and trade in carp became major industries in areas such as Illinois (Bartlett, 1901, 1905). Later, carp began to be regarded as pests, because they destroy aquatic vegetation, stir up sediments and have detrimental effects on water fowl and game fish populations (Moyle, 1984; Marx, 1980). Control measures such as poisoning large bodies of water or reducing water levels immediately after spawning were attempted at considerable expense, but did not generate sustained control. Moyle (1984) recommends that we

> learn to live with and enjoy the carp, as it will always be with us.

Australia and New Zealand

Carp were introduced to Victoria in the latter half of the 19th century. They are thought to harm native fish and water fowl populations (Anon., 1976), as in the USA. Control measures considered included production and distribution of a specific anti-carp virus (Wharton, 1979; McLaren, 1980). Genetic control measures were also suggested but considered unlikely to be effective (Brown, 1980). Legislation was enacted in Victoria against possessing, growing and transporting these ‘noxious fish’, but

> any person taking a noxious fish who immediately kills the fish shall not be liable to any penalty. (Anon., 1976).

Possession of carp is also against the law in New Zealand, where a

dawn swoop uncovers illicit fishery ring

consisting largely of ornamental carp, was reported on the front page of the daily press (Murphy, 1987). A divergent view regards carp as a major asset to Australian fisheries, since they serve as prey for native carnivorous fish like Murray cod (*Maccullochella peeli*) (McLaren, 1980).

South Africa

Carp were introduced to South Africa from England in 1896 and again from Germany in 1955. In 1961, the Dinkelsbühl variety of Aischgrund carp was introduced (Lombard, 1961). The aim of these introductions was to develop a fishery in areas too cold for tilapias. Growth of Aischgrund and Dor-70 varieties and their crossbred were compared in Transkei (Prinsloo & Schoonbee, 1984).

Indonesia

Common carp are grown for human consumption on a large scale. At least 10 different varieties are in commercial use (Pitt, 1984; Sumantadinata & Taniguchi, 1990a,b).

Japan

Common carp were first introduced from China in about 200 AD (McDowell, 1989; Carmines, 1993) and were grown in aquaculture. Different types of ornamental carp (koi) were developed since the 19th century, particularly in the Niigata Prefecture (Carmines, 1993). Chinese, European and local varieties of common carp are present in Japan (Suzuki *et al.*, 1976). Until the mid-1940s they were grown mainly in rice paddies. This was discontinued owing to the widespread use of insecticides toxic to fish (Suzuki, 1979). At present carp are grown largely in irrigation reservoirs and running water ponds.

INTRODUCTION OF CHINESE CARPS

All the important Chinese carps have been introduced to areas outside their endemic distribution in the drainage basins of the major Chinese rivers.

Grass carp

The grass carp has been introduced to almost as many countries as the common carp (Table 7.3a). The main reasons for these introductions

Table 7.3a. Introductions of Grass carp (from Welcomme, 1988).

To	From	Year	Reason	Reproduction
Africa				
Egypt	Hong Kong	1969	A,W	a
Ethiopia	Japan	1975	A,W	a
Kenya	Japan	1969	A,W	a
Mauritius	India	1975	A,W	a
Rwanda	Kenya	1979	A,W	a
S. Africa	Hungary, Germany	1975	A,W	a
Sudan	India	1975	A,W	a
America				
Argentina	Japan	1975	A,W	a
Bolivia	?	1981	A	a
Brazil	Japan	1968, 1979	A,W	a
Colombia	?	?	A	a
Costa Rica	Taiwan	1976	A,W	a
Cuba	USSR	1966, 1976	A,W	a
Honduras	Taiwan	1976	A,W	a
Mexico	China	1965, 1968	A,W	+
Panama	Taiwan	1977, 1978	A,W	a
Peru	Panama	1979	A,W	a
Puerto Rico	USA	1972	A,W	a
USA	Taiwan	1966	W	a, and Mississippi
Asia				
Afghanistan	China	1970s	A, R	a
Bangladesh	Hong Kong	1969	A,W	a
India	Hong Kong	1959, 1968	A,W	a
Indonesia	Malaysia/ Thailand/Japan	1915/49/64	?	a
Korea	Japan	1963	A	a
Malaysia	China	1800s	A	a
Pakistan	China	1964	A	a
Philippines	China	1964	A	a
Sri Lanka	China	1948	?	a
Thailand	China/ Hong Kong	1913	A	a
Vietnam	China	?	A	a

Europe				
Belgium	USSR/Hungary	1967	?	?
Cyprus	Israel	1977	W,S	a
Danube Countries	?	?	?	in Danube system
Denmark	USSR/Malaysia/ Netherlands	1965/66/78	?	a
France	USSR/Hungary Czechoslovakia	1967, 1976	W	a
Germany	Hungary/China	1964, 1970	A,W	a
Greece	Poland	1980	F	a
Hungary	China	1963	A	a, and Danube
Israel	China	?	A	a
Netherlands	Hungary/Taiwan	1966, 1968	W,S	a
Poland	USSR	1965	A	a
Romania	USSR	?	A	a
Sweden	Hungary	1970	W	a
UK	Austria/Hungary/ Germany	1963	W	no
Yugoslavia	Romania/Hungary/ USSR	1963	A,W	a, and Danube
Oceania				
Fiji	Malaysia	1968	A,W	a
Hawaii	Taiwan	1968	A	a
New Zealand	Hong Kong	1967, 1971	W	a

A: aquaculture; a: artificially; W: aquatic weed control; F: fisheries; S: sport.

were aquaculture and the control of aquatic weeds. Artificial reproduction was carried out in most cases. The most common sources of grass carp introductions were China and other Far Eastern countries (Taiwan, Japan,and Hong Kong), though introductions to Europe often came from other European countries such as Hungary and the USSR. Most of these introductions took place in the 1960s and 1970s.

Silver and Bighead carp

Introductions of these fishes took a similar pattern to those of grass carp (Tables 7.3b,c). They were introduced to countries in all the continents, except that bighead carp do not appear to have been introduced to countries in Asia. The main reasons for these introductions were aquaculture. In some cases silver carp were introduced to prevent excessive algal blooms. Strangely enough both species were introduced to Fiji as a source of pituitary glands. As with grass carp, the most common source of introduced fish was China and other countries of the Far East, though European introductions were often from other European countries. Reproduction was artificial in most cases.

Mud and Black carp

Introductions of these fishes were more restricted than those of the other carps (Table 7.3d). Most introductions were to countries in Asia and came from China or Taiwan. Some introductions to Latin American countries also took place. Their aim was aquaculture and, in the case of the black carp introduction to Israel, the control of aquatic molluscs. Reproduction was artifcial in all cases.

SPONTANEOUS SPAWNING OF CHINESE CARPS

It was for long considered that the Chinese carps spawned only in their native Chinese river systems. Artificial propagation, involving hormone treatment, is the only method of producing fry of transplanted Chinese carps. It has replaced the former method of trapping naturally propagated fry in China. A number of cases of natural propagation of Chinese carps, outside their endemic river systems, has been documented (Table 7.4). These include the Tone River in Japan; the Pampangi River in the Philippines; the Volga, Ili, Syr-Dar'ya, Amu-Dar'ya, Kuban, and Terek Rivers in Russia; the Danube; and the Mississippi, Missouri, and Rio Balsas Rivers in America. The report of grass carp fry in a reservoir in Taiwan is probably due to spawning

> in a small feeder stream from which the eggs drifted into the reservoir.
>
> (Stanley, 1976).

Spontaneous spawning has been found most frequently in grass carp, but has been recorded also in silver and bighead carp. We may expect spontaneous spawning in other river systems with hydrological conditions similar to those in the Chinese rivers. Possibly mud and black carp also may spawn spontaneously outside their native rivers.

Table 7.3b. Introductions of Silver carp (from Welcomme, 1988).

To	From	Year	Reason	Reproduction
Africa				
Egypt	Japan	1962	E	a
Ethiopia	Japan	1975	Stocking	a
Madagascar	North Korea	1982	Rice fields	a
Mauritius	India	1976	?	a
Rwanda	Korea	1979	A	a
S. Africa	Israel	1975	A	a
America				
Brazil	Japan/China	1968/79/81	A	a
Costa Rica	Taiwan/China	1976	A	a
Cuba	USSR	1967, 1978	A, F	a
Dominican Republic	Taiwan	1981	A, F	a
Honduras	Taiwan	1976	A	a
Mexico	China	1965	A, Ab	a
Panama	Taiwan	1978	A	a
Peru	Panama	1978	A	a
USA	?	?	Release	?
Asia				
Bangladesh	Japan	1969	A	a
India	Hong Kong/Japan	1959	A	a
Indonesia	Japan/Taiwan	1964, 1969	A	a
Korea	Japan	1963	A	a
Malaysia	China	1800s	A	a
Philippines	China/Taiwan	1964, 1968	A	a
Sri Lanka	China	1948	A	a
Thailand	China/ Hong Kong	1913	A	a
Vietnam	China	?	Aa	
urope				
Belgium	Yugoslavia	1975	Ab	a
Cyprus	Israel	1976	S	a
Czechoslovakia	?	?	S	in Danube
France	Hungary	1975	Ab	a
Germany	Hungary/China	1964/1970	A	a
Greece	Poland	1980	F	a
Hungary	China	1963, 1964	A	in Danube
Israel	?	1966	A	a
Netherlands	Hungary	1966	?	a
Poland	USSR	1965	A	a
Romania	?	?	?	in Danube
Yugoslavia	USSR/Romania/ Hungary	1963	A, Ab	in Danube
Oceania				
Fiji	Malaysia	1968	P	a
New Zealand	?	1969	A	a

A: aquaculture; a: artificially; Ab: aquatic bloom control; E: experimental; F: fisheries; P: pituitaries; S: sport.

Table 7.3c: Introductions of Bighead Carp (from Welcomme, 1988).

To	From	Year	Reason	Reproduction
America				
Brazil	China	1979, 1984	A	a
Costa Rica	Taiwan	1976	A	a
Cuba	USSR	1968, 1976	A	a
Dominican Republic	Taiwan	1981	A, F	a
Mexico	Cuba	1975	A	a
Panama	Taiwan	1978	A	a
Peru	Israel/Panama	1979	A	a
USA	?	?	A	a
Asia				
Indonesia	Japan	1964	A	a
Japan	China	?	A	a
Korea	Taiwan	1963	A	a
Malaysia	China	1800s	A	a
Philippines	Taiwan	1968	A	a
Sri Lanka	China	1948	?	a
Thailand	China	1913	A	a
Vietnam	China	?	A	a
urope				
Danube basin	?	?	A	in Danube
France	Hungary	1975, 1976	A	a
Germany	Hungary	1964	A	a
Hungary	China/USSR	1963/64/68	A	a
Israel	?	1976	A, stocking	a
Italy	Eastern Europe	1975+	S	a
Netherlands	?	1983	in waterways	no
Poland	USSR	1965	A	a
Yugoslavia	Romania/ Hungary/USSR	1963	A	a
Oceania				
Fiji	Malaysia	1968	P	a

A: aquaculture; a: artificially; F: fisheries; P: pituitaries; S: sport

Table 7.3d: Introductions of Mud carp and Black carp. (from Welcomme, 1988).

To	From	Year	Reason	Reproduction
Mud Carp				
America				
Panama	Taiwan	1977	A	+
Asia				
Indonesia	Japan	1964	A	a
Japan	China	?	A	a
Malaysia	China	?	?	a
Thailand	China	?	?a	
Black Carp				
America				
Costa Rica	Taiwan	1979	A	a
Cuba	USSR	1978	A	a
Panama	Taiwan	1978	A	+
Asia				
Thailand	China/ Hong Kong	1913	A	a
Vietnam	China	?	A	a
Europe				
Albania	China	?	A	a
Germany	China	1970	A	a
Israel	?	?	Mollusc control	a

A: aquaculture; a: artificially; +: unknown.

KARYOTYPE

The diploid number of chromosomes in the common carp appears to be 2n=100 (Ojima & Hitotsumachi, 1967; Raicu *et al.*, 1972; Vasilev *et al.*, 1978; Blaxhall, 1983; Rab *et al.*, 1989; Klinkhardt & Buuk, 1990). Earlier reports of 2n=104 (Makino, 1939; Ohno *et al.*, 1967) are apparently due to two pairs of micro-chromosomes not detected in later studies (P.Rab, pers. comm.). Karyotype morphology varies markedly in different studies, irrespective of whether fish of the same or of different stocks were examined (Raicu *et al.*, 1972; Denton, 1973; Blaxhall, 1983; Rab *et al.*, 1989; Klinkhardt & Buuk, 1990). The number of metacentrics appears to vary from 14-28, that of submetacentrics from 6-42, and of subtelocentrics and acrocentrics from 20-62. These large variations may be a result of differences in technique of chromosome preparation, for example direct preparation from tissues or from cell cultures (P.Rab, pers. comm.). Owing to their small size and large number, the definition of each chromosome is somewhat difficult (Blaxhall, 1983).

The cyprinids appear to be divided into two main groups, with diploid numbers of 48-52 in the Chinese carps, and 100 in common carp and *Carassius* spp. (Raicu *et al.*, 1972; Vasilev *et al.*, 1978). The latter

Table 7.4. Spontaneous reproduction of Chinese carps outside their native river systems.

Species	Country	River system	Authority
Grass carp			
	Hungary/ Yugoslavia	Tisza & Danube	Welcomme, 1988
	Japan	Tone	Kuronuma, 1954
	Mexico	Rio Balsas	Stanley, 1976
	Philippines	Pampangi	Stanley, 1976
	Taiwan	Ah Kung Tien Reservoir	Tang, 1960.
	USA	Mississippi	Welcomme, 1988
		Missouri	Brown & Coon, 1991
	USSR	Amu Dar'ya	Stanley, 1976
		Ili	Neszdolly & Mitrofanov, 1975
		Kuban	Stanley, 1976
		Syr-Dar'ya	Verigin *et al.*, 1978
		Terek	Stanley, 1976
		Volga	Stanley, 1976
Silver carp			
	Hungary/Yugoslavia/ Czechoslovakia/ Romania	Danube	Welcomme, 1988
	Japan	Tone	Kuronuma, 1954
	Taiwan	Ah Kung Tien reservoir	Tang, 1960
	USSR	Syr-Dar'ya	Neszdolly & Mitrofanov 1975
Bighead carp			
	USSR	Syr-Dar'ya	Verigin *et al.* 1978
	?	Danube	Welcomme, 1988

are regarded as natural tetraploids (Ohno *et al.*, 1967). In *Carassius auratus langsdorfii*, chromosome number may be 100, 156 or 206 (Kobayasi *et al.*, 1977; Fan & Shen, 1990), that is natural tetraploids, hexaploids or octoploids. Among the diploid chromosome set of common carp, two unattached chromosomes were found in some fish showing morphological characteristics of males. These may be sex chromosomes, indicating that in carp males are the heterogametic sex (Blaxhall, 1983). An unmatched chromosome pair was also detected in a Chinese study referred to by P.Rab (pers. comm.), but not in most other investigations.

GENETIC MARKERS

Scale pattern

The inheritance of scale pattern in common carp has been studied in a number of investigations, and completely defined by Probst (1949a, 1950). The four major scale patterns (Plate 3) are controlled by two alleles, each at two autosomal loci. Wild carp possess the 'scaly' pattern, that is a complete cover of small scales, controlled by the dominant allele at the S locus (SS or Ss). The pattern termed 'mirror' consisting of a partial cover by large mirror scales, is controlled by the recessive allele at this locus (ss). The second locus is lethal when the dominant allele is homozygous. When N is heterozygous (Nn), the scale pattern depends on the allele at the S locus. With the dominant allele (SSNn or SsNn), the scale pattern is termed 'line' or 'linear' and consists of a complete cover of the lateral line by very large and regularly shaped mirror scales. With the recessive allele at the S locus (ssNn), the scale pattern is 'leather' or 'nude', that is the number of mirror scales is very small or they may be completely absent. The number of scales in mirror carp is very variable, ranging from a very small number resembling leather carp, to a complete cover (Kirpichnikov, 1981). The genetic determinants of scale pattern in common carp and *Carassius carassius*, appear to be homologous. F2 generations of crosses between leather carp and (scaly) *C. carassius* segregate into line and fully scaled fish (Lieder, 1957; Kirpichnikov, 1981).

Body coloration

A number of body colorations differing from the wild type are known in common carp. All are based on autosomal recessives. The heredity of 'blue' was described by Probst (1949b) as a simple Mendelian recessive. The characteristics of this colour morph are transparency of the skin, making the blood vessels clearly visible, particularly on the operculum and the bases of the fins. The colour morph 'grey' (Moav & Wohlfarth, 1968) is sometimes difficult to distinguish from the wild type. In our studies we have used blue and grey as a double mark for identifying broodstock (Hines *et al.*, 1974). Another colour morph, 'gold' (Moav & Wohlfarth, 1968; Kirpichnikov, 1981) distinguishable from the wild type at hatching, has also been used for marking broodstock (Moav *et al.*, 1975). Gold is controlled by recessive alleles at two non-linked loci (Katasanov, 1978; Nagy *et al.*, 1979; Cherfas *et al.*, 1992), though in our investigations gold appeared to be controlled by a single recessive allele. Carp possessing both the grey and gold markers are almost colourless, a sort of off-white. In addition to these simple colour markers, yellow, red, blue, black, white and different combinations of these colours appear in ornamental (koi) carp. Investigations on the inheritance of these colours and colour combinations are of recent origin. It appears that some of the single colours, such as white, orange and black, may be true-breeding and appear under the control of single loci, but more complicated patterns, such as 'showa' (red and white areas on a black background) and 'hariwake' (orange areas on a white background) are not true-breeding and their inheritance is not simple (Wohlfarth & Rothbard, 1991; Szweigman *et al.*, 1992). The only documented, visual genetic marker in Chinese carps is albino in grass carp, which is inherited as a Mendelian recessive (Rothbard & Wohlfarth, 1993).

Electrophoretic markers

A number of polymorphic electrophoretic markers has been found, for example transferrin, esterase, malate dehydrogenase, phosphoglucose isomerase, phosphoglucomutase, and lactate dehydrogenase (Valenta *et al.*, 1976; Brody *et al.*, 1976, 1979; Kirpichnikov, 1981). The genetic variability of each marker appears to be under the control of a number of alleles, with a co-dominant relation among the alleles at each locus. The advantage of electrophoretic markers over other known genetic markers in common carp is due to this co-dominant inheritance. It enables the heterozygote to be distinguished from the homozygote for each allelic pair. Uses of these markers as breeding tools are described below.

BIOLOGICAL CHARACTERISTICS OF COMMON CARP

Sex and fertilization

Common carp are normally bisexual, though hermaphrodite individuals occur occasionally (Steffens, 1958) and may even be used for self-fertilization (Kossman, 1971). Hermaphroditism may also be induced, when sex steroids are applied at temperatures sub-optimal for sex-reversal (20°C) (Nagy *et al.*, 1981). Self-fertilization, the most extreme form of inbreeding, does not normally take place in higher animals. It induces a more rapid rate of inbreeding than either sib mating or polar body gynogenesis (Nagy & Csanyi, 1982; see below) and may be a viable alternative to gynogenesis. It is also possible to integrate these two processes (Nagy & Csanyi, 1982). Fertilization in common carp is external, as in most other fishes of importance in aquaculture. This simplifies application of techniques involving treatment of gametes, chromosome manipulation, and genetic engineering.

Fecundity

The common carp is very fecund. A ripe female may lay 120,000 eggs per kilogramme body weight (Bishai *et al.*, 1974; Tomita *et al.*, 1980). The carp's high fecundity has been known for a long time:

> With a few milters and spawners, two or three of each, you may stock a county. (North, 1835).

This has a number of implications. High selection intensities may be applied, performance tests may be carried out with a large number of individuals, and the number of broodstock required is relatively small, which may lead to rapid inbreeding (Falconer, 1960), especially on small farms. Expressing fecundity as number of eggs per kilogramme body weight implies a positive correlation between number of eggs and weight of parent, that is, between fecundity and growth. This means that natural selection, acting directly on fecundity may generate a correlated response on growth. Therefore growth may be close to a selection plateau if fecundity is at a selection plateau as expected for a fertility trait. This is presumably also the case with other fish of high fecundity.

Sex determination and control

The genetic mechanism of sex determination in common carp appears to involve female homogamety (XX) and male heterogamety (XY). Gynogenetically produced progenies are generally all female (e.g., Nagy *et al.*, 1978), but gynogenetic males have been found occasionally among both meiogynes and mitogynes (Komen *et al.*, 1988). This is apparently due to a recessive sex-determining gene (aa), which when homozygous induces maleness, irrespective of the sex chromosome complement, that is, XXaa is male. It is not known whether this locus is on a sex chromosome or an autosome. Androgenesis has been induced in common carp (Grunina *et al.*, 1991; S.Rothbard, pers. comm.), and is expected to generate XX females and YY males, if the latter are viable. These males would be extremely useful in investigations on sex determination and may enable generating all male broods. Blaxhall's (1983) cytological evidence for male heterogamety has been mentioned above. Hormonal sex reversal was generated in carp with the aid of methyl testosterone incorporated at 100μg/g into the feed (Nagy *et al.*, 1981; Gomelskii, 1985). These investigations were carried out on all-female, gynogenetic broods. When the sex-inverted males (but genotypic females) were mated to ordinary females, the progeny were all female (Nagy *et al.*, 1981). This is further evidence for female homogamety in common carp.

Competition within populations

Frequency distributions of length of common carp are asymmetrical and skewed to the right, at stages after yolk sac absorption has been completed. In such populations some individuals, termed shoot carp, are much larger than the population mean. This is a result of competition for food between individuals in the population (Nakamura & Kasahara, 1955, 1956, 1957, 1961). The intensity of skewing (as estimated by the 3rd moment coefficient of asymmetry) increases with increasing stocking density and decreasing amount of food available per individual. Under extreme conditions, which may occur in crowded fry ponds, the smaller individuals may be unable to compete or grow, and the larger ones may become cannibalistic (Kostomarov & Hrabe, 1943; van Damme *et al.*, 1989). Removal of the smaller individuals by the larger ones, thereby providing extra, high quality food for them, is instrumental in increasing the asymmetry of the frequency distribution at both ends. This phenomenon may decrease the likelihood of response to mass selection for faster growth. Selection of the largest individuals may generate a response to food gathering ability under conditions of extreme competition for food, rather than growth rate in commercial management.

Longevity

The rumoured ability of common carp to reach an age of 100 years or more is not supported by any real records (Sarig, 1966). The maximum recorded age of carp from reservoirs, ponds, or aquaria is about 40 years (Alikhuni, 1966; Sarig, 1966; Hinton, 1962), and maximum recorded age under natural conditions does not exceed 16 years (Alikhuni, 1966). This longevity, together with the carp's high fecundity, means that spawners may produce a very large number of progeny. A considerable effort in the genetic improvement of each spawner is therefore justified.

METHODS OF GENETIC IMPROVEMENT

> There are two ways in which the action of the breeder can change the genetic properties of the population; the first by choice of individuals to be used as parents, which constitutes selection, and the second by control of the way in which the parents are mated, which embraces inbreeding and crossbreeding. (Falconer, 1960)

Selection implies choosing some individuals as broodstock for generating the next generation, and excluding others from participation. Selection methods differ according to the method of choosing preferred individuals. Genetic improvement by manipulating the mating system generally involves increasing the rate of crossbreeding above that expected in a random breeding population, with the aim of avoiding inbreeding depression or (which is the same) generating heterosis.

Traditional methods of genetic improvement

Selection of common carp, traditionally carried out in Europe while harvesting 3-year-old fish, consists of selecting the largest fish with some consideration of appearance - a high height: length ratio, a relatively small number of scales in mirror carp, colour, or absence of infectious dropsy scars (Willer, 1933; Mann, 1961; Hofmann, 1975). Several suggestions were made to improve this system and test its results. Phenotypic selection was criticized by Probst (1938a) and Kostomarov (1943), who suggested progeny testing as a more rational method of selective improvement. Schäperclaus (1956) suggested a sequential selection programme according to descent of parents; body shape, scale pattern, and colour; growth and disease tolerance; and performance of progeny. Probst (1938a) enumerated and demonstrated experimental techniques required for selective improvement, including methods of marking fish, artificial propagation, and performance testing. He suggested carrying out diallel crosses and performance testing by stocking all test groups in the same pond. This advanced treatise of aims and methods of selective breeding was not followed by any change in methods employed for over two decades.

Mass Selection

Mass selection involves a choice of individuals according to their phenotypes - the largest when selecting for fast growth, or the survivors after a disease when selecting for disease tolerance. Monitoring the response to mass selection involves the parental generation, in which the selection was carried out, and the progenies of these selected individuals. A random sample may act as control in the parent generation, and its offspring in the progeny generation. The difference in performance between the selected sample and the control in the parental generation is called the selection differential (S), and that in the progeny generation the selection response (R). Effectiveness of selection is estimated by R/S, termed the realized heritability (h^2). The advantages of mass selection are its simplicity, the ability to be carried out with a minimum of facilities and the high possible selection intensities. It is not effective at low realized heritabilities. A long-term mass selection experiment, involving 5 consecutive cycles of bi-directional selection was carried out with common carp (Moav & Wohlfarth, 1976). There was a positive response to down-selection during the first 3 generations, equivalent to a realized heritability of 0.3, and a negative response to up-selection during the first 2 generations, followed by a positive response. The fastest growth of all groups tested was a standard crossbred. A controlled mass selection experiment for height:length generated a response in both directions, equivalent to realized heritabilities of 0.3-0.5. Differences in body shape between different European carp races (Willer, 1933; Hofmann, 1975) may be a result of different intensities of selection practised in the past. It has been assumed that weight and height:length ratio are correlated in common carp, and that selecting for a high ratio was an indirect way of selecting for weight. Empirical estimation showed the correlation was low and insignificant (Stegman, 1968). Owing to the simplicity of mass selection, other selection methods are applied only when response to mass selection is low or when mass selection is not applicable.

Family selection

All types of family selection are difficult to apply and generate low selection intensities, relative to mass selection. Family selection

> is to be preferred when the character selected has a low heritability. The environmental deviations of the individuals tend to cancel each other out in the mean value of the family, and the phenotypic mean of the family comes close to being a measure of its genotypic mean. (Falconer, 1960).

In between-family selection the criterion for selecting or rejecting whole families is mean family performance. In within-family selection the criterion is the deviation of each individual from his family mean, the better individuals of each family being selected. Combined selection may also be carried out. Some family selection has been performed with common carp

(Moav & Wohlfarth, 1976; Nagy *et al.*, 1980; Brody *et al.*, 1981), with apparently promising results. Family selection is the major method in the large scale Norwegian selective breeding effort with rainbow trout (*Oncorhynchus mykiss* Walbaum) and Atlantic salmon (*Salmo salar* L.) (Gjedrem & Skjervold, 1978).

Selection among strains

This is similar to family selection, but uses existing strains instead of establishing families. Criteria for selecting or rejecting strains are the mean strain performance in traits of economic importance. The following are some examples of strains of common carp.

Central Europe

Four major carp races: Aischgrund, Galician, Lausitz and Frankish (Willer, 1933; Hofmann, 1975) have disappeared largely as genetic entities during the period of the two world wars, and carp in central Europe appear to be derivatives of Galician x Lausitz (Komen, 1990).

Russia

Some strains are products of selective breeding schemes (Kirpichnikov, 1981). Ukrainian carp are characterised by their fast growth and high fecundity, particularly under favourable conditions. Ropsha carp were bred specifically for tolerance to low temperatures (Kirpichnikov, 1981). Krasnodar carp were bred for tolerance to infectious dropsy (Kirpichnikov *et al.*, 1993).

Japan

Available strains include a wild one, originally collected from a lake, two domestic Japanese strains, two strains introduced from Germany in 1968, and Chinese Big Belly carp, introduced from Taiwan in 1970 (Suzuki & Yamaguchi, 1980a,b). In comparative growth tests German mirror carp were the fastest and Japanese wild carp the slowest. There is no evidence that this information is applied to choice of commercial strains.

Indonesia

Domestication of common carp in Indonesia began about 150 years ago (Pitt, 1984; Sumantadinata & Taniguchi, 1990) and present strains include Mayalaya (the fastest growing); Sinyonya and Super (intermediate in growth); a Taiwanese strain (presumably Big Belly carp, introduced in 1970); Punten (a product of selective breeding during the 1930s at Punten research station); and several others. There is no evidence for selection among strains.

Israel

Available strains include Dor-70 (Wohlfarth *et al.*, 1980); Nasice (introduced from Yugoslavia in 1970); and Big Belly carp (introduced from Taiwan in 1970) (Moav *et al.*, 1975). Big Belly carp showed very slow relative growth under all but the worst of conditions; the Nasice strain also grew rather badly, apparently as a result of inbreeding depression; whereas Dor-70 grew relatively fast under most conditions (Moav *et al.*, 1975). This information is not used directly for choice of commercial strains, since inter-strain crosses (Dor-70 x Nasice) are stocked as standard into commercial ponds.

Crossbreeding

Systematic crossbreeding involves planned crosses among unrelated individuals or groups, such as among strains of common carp.

> Inbreeding causes an increase in frequency of homozygous genotypes and a decrease of heterozygous genotypes (Falconer, 1960)

and the reverse is the case for crossbreeding. The effects of inbreeding and crossbreeding on traits of economic importance depend on the frequency and dominance relationships of genetic determinants of these traits.

> If the genes that increase the value of the character are dominant over the alleles that reduce the value, then inbreeding will result in a reduction of the population mean. (Falconer, 1960)

Owing to their high fecundity, often resulting in a small number of broodstock, common carp tend to be inbred in captivity. Inbreeding in carp has been shown to reduce growth (Wohlfarth & Moav, 1971; Merla, 1972). A single generation of sib mating may reduce growth by 10-20%. Several generations of uncontrolled mating in captivity, during which some degree of inbreeding is likely to occur, may have a large cumulative effect on growth and survival. Heterosis of growth has been demonstrated repeatedly in crosses among strains of European carp (Andryasheva, 1971; Smisek, 1979, 1981; Bakos, 1979; Berscenyi & Nagy, 1986); in Chinese carp (Zhongbo *et al.*, 1981); in Chinese x European carp (Wohlfarth *et al.*, 1983, 1986); and in European x Japanese carp (Suzuki, 1979; Suzuki & Yamaguchi, 1980a; Shimma *et al.*, 1983; Shimma & Maeda, 1986). In a test involving different isolates of European carp, crossbreds were compared with their parents as both young of the year and one-year-olds. Heterosis of growth was more evident in younger fish (Wohlfarth *et al.*, 1984). Similarly in crosses among cultivated and wild strains of Russian origin, heterosis of growth and survival was most evident during the first month and under harsh conditions (Andryasheva, 1970). This may explain the success of crossbred derivatives for tolerance to low temperatures or manure utilization (see below). Heterosis has also been shown for survival (Andryasheva, 1971; Bakos, 1979; Smisek, 1979, 1981) and tolerance to specific diseases (Kirpichnikov, 1957, 1981; Bauer, 1971; Hines *et al.*,

1974; Sovenyi *et al.*, 1988). A long-term programme, comparing crossbred progenies to their parental lines, showed that heterosis was found commonly among young of the year fish during the summer, but much less during the winter and among adult carp (Wohlfarth, 1993). Heterosis also appeared to be a negative function of growth conditions, being more common when growth was slow. It was found frequently among crossbreds between relatively slow growing lines, but only rarely when Dor-70 was one of the parental lines. Other fast growing lines of common carp may also show growth rates as fast as that of their crossbred progenies.

Breeding for adaptation to unfavourable environments

In northeastern Russia low temperatures and a short growing season severely limit carp cultivation. A carp breed specifically adapted to grow in these conditions was produced by crossing the Galician strain of domestic carp with a cold-tolerant strain of wild carp from the Amur River. This was followed by a series of backcrosses and intercrosses, eventually producing a strain called Ropsha carp (Kirpichnikov, 1981). Crossbred progenies among different groups of Ropsha carp are stocked into ponds in Siberia. There they show better growth and survival than the Galician strain (Romanova *et al.*, 1970; Zonova & Shatnova, 1970), and are more resistant to infectious dropsy and air-bladder disease (Bauer, 1971).

In principle, a similar technique was used to breed carp with fast growth, when manure was used as the principal nutrient. Chinese Big Belly carp are well adapted to these conditions, but grow slowly. Big Belly was crossed with European strains in order to combine their fast growth with Big Belly's ability to survive and graze well under manure utilization. Subsequent comparisons between this crossbred and its parent strains showed that under the best conditions European strains show the fastest growth. In very poor conditions, Big Belly may grow faster than European strains, but in a wide range of intermediate conditions the crossbred is heterotic. The advantage of the crossbred is particularly emphasised when manure is the principal nutrient (Wohlfarth *et al.*, 1983, 1986). Its advantage over truebreeding Big Belly carp has been demonstrated also in Hong Kong (Sin, 1982). Big Belly and its crossbred with European carp is not adapted to grow in cages, where grazing is not possible and the fish subsist on supplemental feed (Wohlfarth & Moav, 1990).

Breeding for disease resistance

Resistance to a number of diseases has been improved by different selection methods. The relative resistance of the Ropsha carp to infectious dropsy is apparently due to the resistance of its River Amur wild strain ancestor (Bauer, 1971). Following artificial infection, mass selection within Ropsha carp populations resulted in further improvement in tolerance (Kirpichnikov *et al.*, 1979, 1987).

Resistance to infectious dropsy was also improved in a direct, long-term selection programme (Kirpichnikov, 1993). Selection was carried out independently in three stocks of common carp during eight generations, and resulted in three selection lines with improved resistance. Crossbreds between these three lines are widely grown in the Caucasus, where this selection programme was carried out, and are called Krasnodar carp. Disease resistance has also been improved by crossbreeding. A crossbred between a Hungarian stock and a Japanese breed of ornamental carp was more resistant than its parental strains to erythrodermatitis (Sovenyi *et al.*, 1988). Similarly, a crossbred between a strain sensitive to epidermal epithelioma and a strain sensitive to swim-bladder inflammation was resistant to both these diseases (Hines *et al.*, 1974).

Breeding to reduce the number of intermuscular bones

Common carp, like many other cyprinids possess a set of small intermuscular bones, called Graten in German, not attached to the skeleton. On average there are 100 of these bones in common and Chinese carp. They detract from the value of the fish by making consumption more difficult, especially for children. Von Sengbusch (1963, 1967) suggested attempting to decrease the number of these bones (possibly to zero) by selection. The first investigation showed a phenotypic range of 70-135 Graten (von Sengbusch & Meske, 1967). In a further study the mean number of Graten was estimated at 86-97 in 8 full sib families (Kossman, 1972). In a genetic survey carried out in Israel, the range in number of Graten was much smaller (Moav *et al.*, 1975b). The survey involved a number of European breeds and a representative of the Big Belly carp, whereas the German study was based on a single north German isolate. The possible reasons for this contradiction are not clear, but they may be connected with the different propagation methods. In the German study, single pair spawns were carried out, producing full sib families, which maximizes genetic variation among groups. In the Israeli study, test groups were produced from multiple spawns, involving a number of parents of each sex.

Chromosome manipulations

This term includes gynogenesis (female parthenogenesis), androgenesis (male parthenogenesis), and polyploidy (triploidy in most cases). Gynogenesis - reproduction without the intervention of the male parent - is a natural method of reproduction in several fish species (Cherfas, 1981; Stein & Geldhauser, 1992). Induction of gynogenesis involves inseminating eggs with sperm inactivated by radioactive or UV irrradiation and restoring diploidy by thermal or pressure shocks. In the complementary process, androgenesis, genetically inactivated ova are fertilized by intact sperm, followed by similar diploid restoration. Polyploidy is induced by shocking normally fertilized eggs. When

shocks are applied shortly after insemination, the second meiotic division is inhibited, resulting in retention of the second polar body. This generates partially heterozygous, diploid gynogens, termed meiogens, or triploids, with irradiated or normal sperm respectively. This method of inducing gynogenesis reduces frequency of heterozygosity, but cannot produce complete homozygosity due to genetic recombination between non-sister chromatids, especially at loci distal to the centromere. Gynogenetic reproduction causes a rate of inbreeding faster than sib mating, during a few first generations only (Nagy & Csanyi, 1982, 1984). Alternating generations of polar body gynogenesis and sib mating is more effective than either system on its own. When shocks are applied later, after the second polar body has been extruded, diploidy may be restored by intervening in the first mitosis. Preventing first cleavage generates completely homozygous individuals termed mitogens, when eggs are inseminated by inactivated sperm. With normal sperm, tetraploids are produced. In common carp, the aims of inducing chromosome manipulations are: production of pure lines for crossbreeding programmes (Nagy et al., 1984); all female progeny - since female carp grow faster than males; sterile triploids for population control; and reduction of variability in colour pattern in ornamental carp (Kawahata, 1989).
In most investigations diploidy was restored by retaining the second polar body. Production of mitogens or tetraploids by inhibiting first cleavage is much more difficult than arresting second polar body extrusion (Nagy, 1987). Gynogenesis or triploidy have been induced in grass carp (Cassani & Caton, 1985; Allen *et al.*, 1986), silver carp (Anon., 1983a; Mizra & Shelton, 1988), bighead carp (Aldridge *et al.*, 1990), and mud carp (Yu *et al.,* 1987). Chromosome manipulation has found some practical applications with rainbow trout and grass carp, but so far not with any of the other carps.

Genetic engineering

This sophisticated genetic technique consists of transferring specific genetic determinants to the target animal from another animal. It requires isolating and cloning a large number of copies of the specific DNA because the chance of each to be integrated into the host genome is very low. Specific DNA sequences of common carp have been cloned (Takeshita *et al.*, 1984; Koren *et al.*, 1989). Growth hormone DNA from various sources has been transferred to target fish, including the transfer of rainbow trout DNA to common carp (Anon., 1988a). Other traits considered for improvement include tolerance to specific diseases and low temperatures (Anon., 1988b). Gene-transfer usually consists of micro-injecting the foreign DNA followed by monitoring its integration into the host genome, reproduction and expression. No advantage of transgenic carp in aquaculture has yet been demonstrated.

Use of electrophoretic markers

Genetic markers are potentially very useful tools in selective breeding investigations. They may be utilized for marking broodstock in order to maintain line purity, and for constructing sophisticated breeding plans, such as Reciprocal Recurrent Selection. They also provide a reliable method of performance testing, since test groups marked by electrophoretic markers may inhabit the same environment during the entire testing programme. With mechanical, rather than genetic marking, each test group is spawned and nursed in a separate facility, until large enough for marking (Moav *et al.*, 1976; Brody *et al.*, 1976). This means that with mechanical marking, at least some of the differences among test groups, isolated by the performance test, may be due to pre-experimental, environmental differences. Little use has been made of electrophoretic markers as a tool in selective breeding. This may be due to the need for individual electrophoretic examinations and weighings of a large number of fish, making this a highly labour-intensive method.

Comparative testing of different genetic groups

A specific and original technique for genetic testing has been developed with common carp. It consists of stocking the different groups to be tested into the same pond, which enables performance testing with a much smaller number of ponds. We termed the co-stocked ponds 'communal ponds' and the method 'communal testing' (Wohlfarth & Moav, 1985). Application of this technique to practical performance testing required a reliable method of differential marking, to distinguish the different groups at the end of the test (Moav *et al.*, 1960a,b). It also required a series of methodological investigations, to monitor its reliability. The genetic correlation between growth in communal and separate testing was found to be unity, meaning that relative performance in communal testing is a reliable estimate of relative performance in separate testing. The genetic regression of growth in communal ponds was estimated as 2, meaning that a given growth difference in communal testing is inflated to twice its value in separate testing (Moav & Wohlfarth, 1974). Different genetic groups are often stocked into the communal testing ponds at differing initial weights owing to incomplete control over conditions in nursery ponds. This is a source of biassing growth data, since initial weight and weight gain are positively correlated (Wohlfarth & Moav, 1972). Initially this bias was corrected with the aid of a factor empirically derived in each communal test. It involved the 'multiple nursing technique', specifically constructed for this purpose (Wohlfarth & Moav, 1985). Later a large body of accumulated data enabled computing a multiple regression equation, equating the predicted correction factor (b) to the mean initial weight (X) and mean weight gain (Y) of all genetic groups in a particular test. This equation was:

$$b = 4.54 - 0.113X + 0.0671Y \quad \text{(Wohlfarth \& Milstein, 1987).}$$

Its use reduced the effort and facilities required for empirical estimation. Communal testing enables any pond to be used as a (communal) testing pond, including farm ponds not situated in a research establishment. There is no requirement for a similar management in replicated communal ponds, a vital condition in separate testing. Communal testing is also used with other fishes in aquaculture, but without carrying out similar methodological investigations.

SUMMARY

The common carp was probably the first fish in aquaculture. Some stocks are domesticated, and some selective breeding schemes have been practised for generations. This is probably a result of the common carp's long association with humankind, its ease of spawning in captivity and its longevity. From its origin in central Eurasia, carp spread naturally east and west, and was distributed by man the world over, wherever conditions, particularly temperature, permitted its growth and reproduction. The common carp is an important component of aquaculture in the Far East and eastern Europe. In the USA and southern Australia it is regarded as a pest, and efforts, invariably unsuccessful, are made to eradicate it, or at least to contain the population. The carp's high fecundity and longevity make it a very suitable candidate for selective improvement. Its external fertilization enables artificial propagation and simplifies advanced genetic techniques. Domesticated carp breeds differ from wild carp in traits such as body shape, growth rate, sexual maturity, and behaviour. This is likely to be the result of the domestication process and some mass selection by which, presumably, it was accompanied. Mass selection can be applied successfully to alter body shape and apparently also to improve disease resistance, but does not generate a response in growth rate. Heterosis of growth rate, survival, and disease resistance has been demonstrated repeatedly. Selection among existing strains, either as a selection method in itself, or as a preliminary to further improvement schemes, has hardly been applied. A project of selective breeding, aimed at reducing the number of intermuscular bones, was abandoned because of inconsistent results. Investigations on including chromosome manipulations and the production of transgenic fish in common carp have not been applied to practical aquaculture. A method for the genetic testing of common carp was developed, which enables carrying out large scale performance testing with an economy of facilities.

The vast yield of Chinese carps makes them promising subjects for investigations into their genetic management, but few studies have been carried out. Domestication must have begun, since fry for stocking aquaculture facilities are produced largely by captive spawning. As far as we can tell this has not been accompanied by any improvement in their performance. On the contrary, the small amount of available information points to a deterioration in performance of domesticated stocks of silver and bighead carp. Chromosome manipulations have been induced in some of the Chinese carps, but, as with common carp, have not been applied to their genetic improvement.

REFERENCES

Aldridge, F.J., Marston, R.Q. & Shireman, J.V. 1990. Induced triploids and tetraploids in bighead carp *Hypophthalmichthys nobilis*, verified by multi-embryo cytofluorimetric analysis. *Aquaculture* **87**, 121-131.

Alikhuni, K.H. 1966. Synopsis of Biological Data on Common carp *Cyprinus carpio* Linnaeus 1758 (Asia and the Far East). *FAO Fisheries Synopsis* **31.1**.

Allen, S.K., Thiery, R.G. & Hagstrom, N.T. 1986. Cytological evaluation of the likelihood that triploid grass carp will reproduce. *Transactions of the American Fisheries Society* **115**, 841-848.

Andryasheva, M.A. 1970. Heterosis in intraspecific crossings of carp. In *Selective Breeding of Carp and Intensification of fish breeding in ponds*. (Kirpichnikov, V.S. ed.) pp.56-73. Springfield, VA: U.S.Dept. of Commerce.

Andryasheva, M.A. 1971. Commercial hybridisation and heterosis in fish culture. In *Seminar/Study Tour in the USSR on Genetic Selection and Hybridisation of Cultivated Fishes*. pp.248-262. Rome: FAO/UNDP.

Anon. 1976. A proposal to assess the impact of European carp on fish and waterfowl. Victoria: Fisheries & Wildlife Division, Ministry of Conservation.

Anon. 1983a. Gynogenetic catla and silver carp produced. *CIFRI Newsletter* **6(1)**, 2.

Anon. 1983b. Siberian carp success. *Fish Farming International* **10(5)**, 14.

Anon. 1988a. Scientists insert gene into carp to speed their growth. *The Baltimore Sun*.

Anon. 1988b. Transgenics - study turns to fish. *Fish Farmer* **11(4)**, 25.

Babushkin, U.P. 1987. The selection of a winter resistant carp. In *Selection, Hybridization and Genetic Engineering in Aquaculture* (Tiews, K. ed.). Vol.I, pp.447-454. Berlin: Heenemann.

Bakos, J. 1979. Crossbreeding Hungarian races of the Common Carp to develop more productive hybrids. In *Advances in Aquaculture* (Pillay, T.V.R & Dill, W. eds.) pp.633-635. Surrey: Fishing News Books.

Balon, E.K. 1974. Domestication of the carp *Cyprinus carpio* L. Toronto, Canada: Royal Ontario Museum, Life Sciences Miscellaneous Publication.

Bartlett, S.P. 1901. Discussion on carp. *Transactions of the American Fisheries Society* **30**, 114-132.

Bartlett, S.P. 1905. Carp, as seen by a friend. *Transactions of the American Fisheries Society* **34**, 207-216.

Bauer, O.N. 1971. Control of fish diseases by selection. In *Seminar/Study Tour in the USSR on Genetic Selection and Hybridisation of Cultivated Fishes*. pp.335-342. Rome: FAO/UNDP.

Berka, R. 1985. A brief insight into the history of Bohemian carp pond management. In *Aquaculture of Cyprinids*. (Billard, R. & Marcel, J. eds.) pp.35-40. Paris: INRA.

Berscenyi, M. & Nagy, A. 1986. Predicting success of hybridisation programs in Hungary. In *Third World Congress on Genetics Applied to Livestock Production.* pp. 417-422. Lincoln NE.

Bishai, H.M., Isaak, M.M. & Labib, W. 1974. Fecundity of the mirror carp *Cyprinus carpio* L. at the Serow fish farm (Egypt). *Aquaculture* **4**, 257-265.

Blaxhall, P.C. 1983. Chromosome karyotyping of fish, using conventional and G-banding methods. *Journal of Fish Biology* **22**, 417-424.

Brody, T., Wohlfarth, G., Hulata, G. & Moav, R. 1981. Application of electrophoretic genetic markers to fish breeding. IV. Assessment of breeding value of full sib families. *Aquaculture* **24**, 175-186.

Brody, T., Moav, R., Abramson, Z.V., Hulata, G. & Wohlfarth, G. 1976. Application of electrophoretic genetic markers to fish breeding. II: Genetic variation within maternal half-sibs in carp. *Aquaculture* **9**, 351-365.

Brody, T., Kirsht, D., Parag, G., Wohlfarth, G., Hulata, G. & Moav, R. 1979. Biochemical genetic comparison of the Chinese and European races of the common carp. *Animal Blood Groups Biochemical Genetics* **10**, 141-149.

Brown, A.M. 1980. Carp program. An evaluation of the role of genetics in the management of Victorian populations of carp (*Cyprinus carpio* L.). Victoria: Fisheries & Wildlife Division, Ministry for Conservation.

Brown, D.J. & Coon, T.G. 1991. Grass carp larvae in the lower Missouri River and its tributaries. *North American Journal of Fisheries Management* **11**, 62-66.

Carmines, C. 1993. Brief history of nishikigoi developments. *Koi USA* **17 (6)**, 52.

Cassani, J.R. & Caton, W.E. 1985. Induced triploidy in grass carp, *Ctenopharyngodon idella* Val. *Aquaculture* **46**, 37-44.

Cherfas, N.B. 1981. Gynogenesis in fishes. In *Genetic Bases of Fish Selection* (Kirpichnikov, V. S. ed.). pp.255-273. Berlin: Springer.

Cherfas, N.B., Peretz, Y. & Ben Dom, N. 1992. Inheritance of the orange type pigmentation in Japanese carp (koi) in the Israeli stock. *Israel Journal of Aquaculture Bamidgeh* **44**, 32-34.

Committee for the Collection of Experiences on the Cultivation of Freshwater Fish Species in China. 1981. Science of the cultivation of freshwater fish in China. Ottawa, Ontario: IDRC Transl. **TS16c**.

Cooper, L.(ed.). 1987. *Carp in North America*. Bethesda, Maryland: American Fisheries Society.

Damme, P. van, Appelbaum, S. & Hecht, T. 1989. Sibling cannibalism in koi carp, *Cyprinus carpio* L., larvae and juveniles reared under controlled conditions. *Journal of Fish Biology* **34**, 855-863.

Denton, T.E. 1973. *Fish chromosome methodology*. Springfield, Illinois: Charles C.Thomas.

Falconer, D.S. 1960. *Introduction to Quantitative Genetics*. Edinburgh: Oliver & Boyd.

Fan, Z. & Shen, J. 1990. Studies on the evolution of bisexual reproduction in crucian carp (*Carassius auratus gibelio* Bloch). *Aquaculture* **84**, 235-244.

Gjedrem, T. & Skjervold, H. 1978. Improving salmon and trout farm yields through genetics. *World Review of Animal Production* **14(3)**, 29-38.

Gomelskii, B.I. 1985. Hormonal sex inversion in the carp (*Cyprinus carpio* L.). *Ontogenez* **16**, 398-405. (In Russian, with English summary).

Grunina, A.S., Gomelskii, B.I. & Nayfakh, A.A. 1991. Induction of diploid androgenesis in common carp and production of androgenesis hybrids between common carp and crucian carp. Poster presented at 4th International Symposium on Genetics in Aquaculture, Wuhan. China. 29 April - 3 May 1991.

Hines, R.S., Wohlfarth, G.W., Moav, R. & Hulata, G. 1974. Genetic differences in susceptibility to two diseases among strains of the common carp. *Aquaculture* **3**, 187-197.

Hinton, S. 1962. Longevity of fishes in captivity, as of September 1956. *Zoologica* **47(2)**, 105-116.

Hoffmann, W.E. 1934. Preliminary notes on the freshwater fish industry in South China, especially Kwangtung Province. *Lignan University Science Bulletin* **5**, 70pp.

Hofmann, J. 1975. *Der Teichwirt. Zucht und Haltung des Karpfens*. 4th ed. Hamburg: Paul Parey.

Huner, J.V. 1985. Common carp is dominant species for warm water aquaculture in France. *Aquaculture Magazine* **11(5)**, 32-40.

Katasanov, V.Ya. 1978. Color hybrids of common and ornamental (Japanese) carp III. Inheritance of blue and orange color types. *Soviet Genetics* **14**, 1522-1528.

Kawahata, A. 1989. Can we copy koi - duplicate color patterns on koi - with biotechnology? *Rinko* **150**, 46-47.

Kirpichnikov, V.S. 1957. Karpfenzucht in Norden der Ud.S.S.R. *Deutsche Fischerei Zeitung* **4**, 213-227.

Kirpichnikov, V.S. 1981. *Genetic Bases of Fish Selection*. New York: Springer-Verlag.

Kirpichnikov, V.S., Ilyasov, Yu.I., Shart, L.A., Vikhman, A.A., Ganchenko, M.V.,

Ostashevsky, A.L., Simonov, V.L., Tikhonov, G.F. & Tyurin, V.V. 1993. Selection of Krasnodar common carp (*Cyprinus carpio* L.) for resistance to dropsy: principal results and prospects. *Aquaculture* **111**, 7-20.

Kirpichnikov, V.S., Faktorovich, K A., Ilyasov, Yu.I. & Shart. L.A. 1979. Selection of common carp (*Cyprinus carpio* L.) for resistance to dropsy. In *Advances in Aquaculture* (Pillay, T.V.R. & Dill, W. eds.) pp.628-633. Farnham: Fishing News Books Ltd.

Kirpichnikov, V.S., Ilyasov, Yu.I. & Shart, L.A. 1987. Herauszuchtung einer Karpfenrasse mit erhöhter Resistenz gegen die infektiose Bauchwassersucht. *Fortschrittliche Fischereiwissenschaft* **5/6**, 113-120.

Klinkhardt, M.B. & Buuk, B. 1990. Die Chromosomen des Karpfens (*Cyprinus carpio*). *Zeitschrift für die Binnenfischerei der DDR* **37**, 188-191.

Kobayasi, H., Nakano, K. & Nakamura, M. 1977. On the hybrids, 4n Ginbuna (*Carassius auratus langsdorfii)* x Kinbuna (*C. auratus* subsp.), and their chromosomes. *Bulletin of the Japanese Society of Scientific Fisheries* **43**, 31-37.

Komen, J. 1990. Clones of common carp. Ph.D. thesis, Wageningen University, The Netherlands.

Komen, J., Duynhouwer, J., Richter, C.J.J. & Huisman, E.A. 1988. Gynogenesis in the common carp (*Cyprinus carpio* L.). I: Effects of genetic manipulation of sexual products and incubation conditions of eggs. *Aquaculture* **69**, 227-239.

Komen, J., Richter, C.J.J. & Huisman, E.A. 1987. The production of gynogenetic inbred lines of common carp, *Cyprinus carpio* L. In *Third International Symposium on Reproductive Physiology of Fish* (Idler, D.R., Crim, L.W. & Walsh, J.D. eds.) p.132. St. John's Newfoundland: Memorial University Press.

Koren, Y., Sarig, S., Ber, R. & Daniel, V. 1989. Carp growth hormone: molecular cloning and sequencing of cDNA. *Gene* **77**, 309-315.

Kossmann, H. 1971. Hermaphroditismus und Autogamie beim Karpfen. *Naturwissenschaften* **58**, 328-329.

Kossmann, H. 1972. Untersuchungen über die genetische Varianz der Zwischenmuskelgraten des Karpfens. *Theoretical and Applied Genetics* **42**, 130-135.

Kostomarov, B. 1943. Uber die züchterische Auswahl in der Karpfenzucht. *Zemedelsko Arkhiv (Brno)* **33(9-10)**, 1-19. (In Czech, with German abstract).

Kostomarov, B. & Hrabe, S. 1943. Die Kannibalismus bei der Karpfenbrut (*Cyprinus carpio* L.). *Archiv für Hydrobiologie* **40**, 265-278.

Kozakova, E.(ed.). 1987. *Czechoslovak Agriculture*. Prague: State Publishing House.

Li, S. 1989. Genetic evaluation of Chinese carps. *Ambio* **19**, 411-415.

Li, S., Weimin, L., Changdie, P. & Zhao, P. 1987a. A genetic study of the growth performance of silver carp from the Chiangjiang and Zhujiang rivers. *Aquaculture* **65**, 93-104.

Li, S., Weimin, L., Changdie, P. & Zhao, P. 1987b. Growth performance of different populations of silver carp and big head. In *Selection, Hybridisation and Genetic Engineering in Aquaculture* (Tiews, K. ed.) Vol I: pp.243-256. Berlin: Heenemann.

Lieder, V. 1957. Die Bewertung der Beschuppung des Karpfens bei der Zuchtauslese. *Deutsche Fischerei Zeitung* **4**, 206-213.

Liu, C.K. 1941. On the spawning ground of carp in Ch'u Ho. *Sinensia* **2**, 227-233.

Lombard, G.L. 1961. The first importation and production of Aischgrund carp in South Africa. *Fauna & Flora (Pretoria)* **12**, 79-85.

MacCrimmon, H.R. 1968. Carp in Canada. *Bulletin of the Fisheries Research Board of Canada* **165**.

Makino, S. 1939. The chromosomes of the carp *Cyprinus carpio* including those of some related species of Cyprinidae for comparison. *Cytologia (Tokyo)* **9**, 430-440.

Mann, H. 1961. Fish cultivation in Europe. In *Fishes as Food* (Bergstrom, G. ed.) pp.70-102. New York: Academic Press.

Marx, W. 1980. Plug ugly minnows or living jewels, carp stir emotions. *Smithsonian Magazine* **11(2)**, 54-63.

McDowell, A.(ed.). 1989. *The Interpret Encyclopaedia of Koi.* London: Salamander Books.

McLaren, P. 1980. Is carp an established asset? *Fisheries* **5(6)**, 31-32.

Merla, G. 1972. Ungünstige Inzüchtfolgen in der Karpfenteichwirtschaft. *Zeitschrift der Binnenfischerei der DDR* **19**, 153-157.

Minghua, L., Junbao, S. & Tieqi, Z. 1992. Selective technique of good breeds in *Cyprinus carpio haematopterus* (Heilongjiang common carp) and *C. carpio* (mirror carp). *Journal of Fisheries of China (Shanghai)* **6**, 7-15. (In Chinese with English summary).Mizra, J.A. & Shelton, W.L. 1988. Induction of gynogenesis and sex reversal in silver carp. *Aquaculture* **68**, 1-14.

Moav, R. & Wohlfarth, G.W. 1968. Genetic improvement of yield in carp. *FAO Fisheries Report* **44(4)**, 12-29.

Moav, R. & Wohlfarth, G.W. 1974. Magnification through competition of genetic differences in yield capacity in carp. *Heredity* **33**, 181-202.

Moav, R. & Wohlfarth, G.W. 1976. Two-way selection for growth rate in the common carp (*Cyprinus carpio* L.) *Genetics* **82**, 83-101.

Moav, R., Brody, T., Wohlfarth, G.W. & Hulata, G. 1976. Application of electrophoretic markers to fish breeding. I: Advantages and methods. *Aquaculture* **9**, 217-228.

Moav, R., Finkel, A. & Wohlfarth, G.W. 1975b. Variability of intermuscular bones, vertebrae, ribs, dorsal fin rays and skeletal disorders in the common carp. *Theoretical and Applied Genetics* **46**, 33-43.

Moav, R., Hulata, G. & Wohlfarth, G.W. 1975a. Genetic differences between the Chinese and European races of the common carp. I. Analysis of genotype-environment interactions for growth rate. *Heredity* **34**, 323-340.

Moav, R., Wohlfarth, G.W. & Lahman, M. 1960a. Genetic improvement of carp II. Marking fish by branding. *Bamidgeh* **12**, 49-53.

Moav, R., Wohlfarth, G.W. & Lahman, M. 1960b. An electric instrument for brandmarking fish. *Bamidgeh* **12**, 92-95.Moyle, P.B. 1984. America's carp. *Natural History* **93(9)**, 42-51.

Müller, S. 1989. Geschichte der Fischzucht in Europa - Teil VII. *Zeitschrift der Binnenfischerei der DDR* **36**, 10-14.

Murphy, T. 1987. Dawn swoop uncovers illicit fishery ring. *The New Zealand Herald (Auckland)*. 9 February 1987.

Nagy, A. 1987. Genetic manipulations performed on warm water fish. In *Selection,*

Hybridization and Genetic Engineering in Aquaculture (Tiews, K. ed.) pp.163-173. Berlin: Heenemann.

Nagy, A. & Csanyi, V. 1978. Utilisation of gynogenesis in genetic analysis and practical animal breeding. In *International Seminar on Increasing the Productivity by Selection and Hybridization* (Olah, J. & Krasznai, Z. eds) pp.16-30. Szarvas, Hungary.

Nagy, A. & Csanyi, V. 1982. Changes in genetic parameters in successive gynogenetic generations and some calculations for carp gynogenesis. *Theoretical and Applied Genetics* **63**, 105-110.

Nagy, A. & Csanyi, V. 1984. A new breeding system using gynogenesis and sex reversal for fast inbreeding in carp. *Theoretical and Applied Genetics* **67**, 485-490.

Nagy, A., Berscenyi, M. & Csanyi, V. 1981. Sex reversal in carp (*Cyprinus carpio*) by oral administration of methyl testosterone. *Canadian Journal of Fisheries and Aquatic Sciences* **38**, 725-728.

Nagy, A., Csanyi, V., Bakos, J. & Berscenyi, M. 1984. Utilisation of gynogenesis and sex reversal in commercial carp breeding: growth of the first gynogenetic hybrids. *Aquacultura Hungarica* **4**, 7-16.

Nagy, A., Csanyi, V., Bakos, J. & Horvath, L. 1980. Development of a short term laboratory system for the evaluation of carp growth in ponds. *Bamidgeh* **33**, 6-15.

Nagy, A., Rajki, K., Bakos, J. & Csanyi, V. 1979. Genetic analysis in carp (*Cyprinus carpio*) using gynogenesis. *Heredity* **43**, 35-40.

Nagy, A., Rajki, K., Horvath, L. & Csanyi, V. 1978. Investigation on carp, *Cyprinus carpio* gynogenesis. *Journal of Fish Biology* **13**, 215-224.

Nakamura, N. & Kasahara, S. 1955. A study of the phenomenon of the tobi koi or shoot carp. I. On the earliest stage at which the shoot carp appears. *Bulletin of the Japanese Society for Scientific Fisheries* **21**, 73-76. (In Japanese with English summary).

Nakamura, N. & Kasahara, S. 1956. A study of the phenomenon of the tobi koi or shoot carp. II. On the effect of particle size and quantity of food. *Bulletin of the Japanese Society for Scientific Fisheries* **21**, 1022-1024. (In Japanese with English summary).

Nakamura, N. & Kasahara, S. 1957. A study of the phenomenon of tobi koi or shoot carp. III. On the results of culturing the modal group and the growth of carp fry reared individually. *Bulletin of the Japanese Society of Scientific Fisheries* **22**, 674-678.

Nakamura, N. & Kasahara, S. 1961. A study of the phenomenon of tobi koi or shoot carp IV. Effects of adding a small number of larger individuals to the experimental batches of carp fry and culturing density upon the occurrence of shoot carp. *Bulletin of the Japanese Society of Scientific Fisheries* **27**, 958-962. (In Japanese with English summary).

Neszdolly, V.K. & Mitrofanov, V.P. 1975. Natural reproduction of the grass carp *Ctenopharyngodon idella* in the Ili River. *Journal of Ichthyology* **15**, 927-933.

North, R. 1835. *A treatise on fish and fish ponds*. London: J.Goodwin.

O'Grady, K.T. & Spillet, P.B. 1985. A comparison of pond culture of carp, *Cyprinus carpio* L., in Britain and mainland Europe. *Journal of Fish Biology* **26**, 701-714.

Ohno, S., Muramoto, J., Christian, L. & Atkin, N.B. 1967. Diploid-tetraploid relationship among old-world members of the fish family Cyprinidae. *Chromosoma* **23**, 1-9.

Ojima, J. & Hitotsumachi, H. 1967. Cytogenetic studies in lower vertebrates. IV. A note on the chromosomes of the carp (*Cyprinus carpio*) in comparison with those of the tuna and the goldfish (*Carassius carassius*). *Japanese Journal of Genetics* **42**, 163-167.

Pitt, R. 1984. *Breeding common carp in Indonesia, hatchery manual and review.* Sumatra Fisheries Development Project. London: Sea Fish Industry Authority.

Prinsloo, J.E. & Schoonbee, H.J. 1984. Evaluation of the relative growth performance of three varieties of the European common carp *Cyprinus carpio*, in Transkei. *Water South Africa* **10**, 105-107.

Probst, E. 1938a. *Neue Wege der Karpfenzuchtung*. Neudamm: Verlag Neumann. (Reprinted from *Fischerei Zeitung* **41**).

Probst, E. 1938b. *Die kunstliche Befruchtung bei Karpfen und Schleien*. Neudamm: Verlag Neumann. (Reprinted from *Fischerei Zeitung* **40**).

Probst, E. 1949a. Vererbungsuntersuchungen beim Karpfen. *Allgemeine Fischerei Zeitung* **74(21)**, 1-8.

Probst, E. 1949b. Der Blauling Karpfen. *Allgemeine Fischerei Zeitung* **74**, 232-238.

Probst, E. 1950. Todesfaktor bei der Vererbung des Schuppenkleides des Karpfens. *Allgemeine Fischerei Zeitung* **75**, 369-370.

Rab, P., Pokorny, J. & Roth, P. 1989. Chromosome study of the common carp, *Cyprinus carpio*. I. Karyotype of Amurian carp, *C. carpio haematopterus. Caryologia* **42**, 27-36.

Raicu, P., Taisescu, E. & Alexandrina, C. 1972. Diploid chromosome complement of the carp. *Cytologia* **37**, 355-358.

Romanova, Z.T., Petrov, E.B. & Zabegalin, N.I. 1970. Evaluation of the fishery qualities of the northern hybrid carp in fish farms of the Pskov region. In *Selective Breeding of Carp and Intensification of Fish Breeding in Ponds* (Kirpichnikov, V.S. ed.) pp.171-183. Washington: National Science Foundation.

Rothbard, S. & Wohlfarth, G.W. 1993. Inheritance of albinism in the grass carp *Ctenopharyngodon idella*. *Aquaculture* **115**, 13-17.

Sarig, S. 1966. Synopsis of the Biological data on common carp *Cyprinus carpio* Linnaeus 1758 (Near East and Europe). *FAO Fisheries Synopsis* **31-2**.

Schäperclaus, W. 1956. Die Bewertung des Karpfens bei der Zuchtauslese. *Zeitschrift für fischerei und deren Hilfswissenschaften* **(NF) 10**, 105-135.

Sengbusch, R. von 1963. Fische ohne Graten. *Der Zuchter* **33**, 284-286.Sengbusch, R. von 1967. Eine Schnellbestimmungsmethode der Zwischenmuskelgraten bei Karpfen zur Auslese von gratenfreien Mutanten. *Der Zuchter* **37**, 275-276.

Sengbusch, R von & Meske, C. 1967. Auf dem Wege zum gratenlosen Karpfen. *Der Zuchter* **37**, 271-274.

Shimma, H. & Maeda, H. 1986. Growth and body composition of F1 hybrid between

Yamoto (female) and mirror (male) carps *Cyprinus carpio*. *Bulletin of the National Research Institute of Aquaculture* **10**, 1-9.

Shimma, H., Suzuki, R. & Yamaguchi, M. 1983. Growth performance and body composition of F1 hybrids between Yamato and mirror carp reared with four kinds of practical diets. *Bulletin of the National Research Institute of Aquaculture* **4**, 1-8.

Sin, W.A. 1982. Stock improvement of the common carp in Hong Kong through hybridisation with the introduced Israeli race Dor-70. *Aquaculture* **29**, 299-304.

Smisek, J. 1979. The hybridization of the Vodnany and Hungarian lines of carp. *Buletin VURH Vodnany* **15(1)**, 3-12.

Smisek, J. 1981. The weight, conformation and resistance of carp fry in hybrid lines. *Buletin VURH Vodnany* **17(4)**, 12-19

Sovenyi, J.F., Berscenyi, M. & Bakos, J. 1988. Comparative examination of susceptibility of two enotypes of carp *Cyprinus carpio* L. to infection with *Aeromonas salmonicida*. *Aquaculture* **70**, 301-308.

Stanley, J.G. 1976. Reproduction of the grass carp (*Ctenopharyngodon idella*) outside its native range. *Fisheries* **1(3)**, 7-10.

Steffens, W. 1958. *Der Karpfen*. Wittenberg: Ziemsen Verlag.

Steffens, W. 1966. Das Domestikationsproblem beim Karpfen. *Verhandlungen des internationalen Verein der Limnologie* **16(3)**, 1441-1448.

Stegman, K. 1968. The estimation of the quality of carp by means of length/height ratio and relative weight gains. *FAO Fisheries Report* **44(4)**, 160-168.

Stein, H. & Geldhauser, F. 1992. Beobachtungen zur Verbreitung des triploiden Giebels (*Carassius auratus*) im Donauraum. *Fischer und Teichwirt* **43**, 291-292.

Sumantadinata, K. & Taniguchi, N. 1990a. Study on morphological variation in Indonesian common carp stocks. *Nippon Suisan Gakkaishi* **56**, 879-886.

Sumantadinata, K. & Taniguchi, N. 1990b. Comparison of electrophoretic allele frequencies and gene variability of common carp stocks from Indonesia and Japan. *Aquaculture* **88**, 263-271.

Suzuki, R. 1979. The culture of common carp in Japan. In *Advances in Aquaculture* (Pillay T.V.R. & Dill, W. eds.) pp.161-166. Farnham: Fishing News Books.

Suzuki, R. & Yamaguchi, M. 1980a. Improvement in quality of common carp by crossbreeding. *Bulletin of the Japanese Society for Scientific Fisheries* **46**. 1427-1434.

Suzuki, R. & Yamaguchi, M. 1980b. Meristic and morphometric characters of five races of *Cyprinus carpio*. *Japanese Journal of Ichthyology* **27**, 199-206.

Suzuki, R., Yamaguchi, M., Ito, T. & Toi, J. 1976. Differences in growth and survival in various races of the common carp. *Bulletin of the Freshwater Fisheries Research Laboratory (Tokyo)* **26**, 59-69.

Szweigman, D., Rothbard, S. & Wohlfarth, G.W. 1992. Further observations on the inheritance of color in koi. *Nichrin 92-1* **293**, 37-41.

Takeshita, S., Aoki, T., Fukumaki, Y. & Takagi, Y. 1984. Cloning and sequence analysis of cDNA for the globin mRNA of carp *Cyprinus carpio*. *Biochimica et Biophysica Acta* **783**, 265-271.

Tang, Y.A. 1960. Reproduction of the Chinese carps *Ctenopharyngodon idella* and *H. molitrix* in a reservoir in Taiwan. *Japanese Journal of Ichthyology* **7**.

Tapiador, D.D., Henderson, H.F., Delmendo, M.N. & Tsutsuy, H. 1977. Freshwater fisheries and aquaculture in China. *FAO Fisheries Technical Papers* **168**, 84pp.

Taverner, J. 1600. *Certaine experiments concerning fish and fruite*. London: William Ponsonby.

Tomita, M., Iwahashi, M. & Suzuki, R. 1980. Number of spawned eggs, ovarian eggs and egg diameter and percent eyed eggs with reference to size of the female carp. *Bulletin of the Japanese Society for Scientific Fisheries* **46**, 1077-1081.

Valenta, M., Stratil, A., Slechtova, V., Kalal, L. & Slechta, V. 1976. Polymorphism of transferrin in carp (*Cyprinus carpio*): determination, isolation and partial characterisation. *Biochemical Genetics* **14**, 27-45.

Vasilev, V.P., Makeeva, A.P. & Rayabov, I.N. 1978. The study of chromosome complexes in cyprinid fish and their hybrids. *Genetika* **14**, 1453-1460.

Verigin, B.V., Makeeva, A.P. & Zaki Mohamed, M.I. 1978. Natural spawning of the silver carp *Hypophthalmichthys molitrix,* the bighead carp *Aristichthys nobilis* and the grass carp *Ctenopharyngodon idella* in the Syr-Dar'ya River. *Journal of Ichthyology* **18**, 143-146.

Wei, Z. & Xinloo, C. 1986. Systematic study of the genus *Cyprinus* (Pisces: Cyprinidae) in Yunnan, China. *Zoological Research* **7**, 297-310. (In Chinese with English abstract).

Welcomme, R.L. 1988. International introductions of inland aquatic species. *FAO Fisheries Technical Papers* **294**, 318pp.Wharton, J.C.F. 1979. Impact of exotic animals, especially European carp *Cyprinus carpio* on native fauna. *Fisheries and Wildlife Paper* **20**, Victoria, Australia.

Willer, A. 1933. Uber Zuchtziele in der Karpfenteichwirtschaft. *Mitteilungen der Deutschen Landeswirtschaftsgesellschaft* **10/11**, 1-3.

Wohlfarth, G.W. 1984. Common carp. In *Evolution of domestic animals* (Mason, I. L. ed.) pp. 375-380. London: Longman.

Wohlfarth, G.W. 1993. Heterosis for growth rate in common carp. *Aquaculture* **113**, 31-46.

Wohlfarth, G.W. & Milstein, A. 1987. Predicting correction factors for differences in initial weight among genetic groups of common carp in communal testing. *Aquaculture* **60**, 13-25.

Wohlfarth, G.W. & Moav, R. 1971. Genetic investigations and breeding methods of carp in Israel. *FAO/UNDP (TA) Report* **2926**, 160-185.

Wohlfarth, G.W. & Moav, R. 1972. The regression of weight gain on initial weight in carp. *Aquaculture* **1**, 7-28.

Wohlfarth, G.W. & Moav, R. 1985. Communal testing, a method of testing the growth of different genetic groups of common carp in earthen ponds. *Aquaculture* **48**, 143-157.

Wohlfarth, G.W. & Moav, R. 1990. Genetic differences between the Chinese and European races of the common carp. 6. Growth of fish in cages. *Theoretical and Applied Genetics* **79**, 693-698.

Wohlfarth, G.W. & Rothbard, S. 1991. Preliminary investigations on color inheritance in Japanese ornamental carp (nishikigoi). *Israeli Journal of Aquaculture Bamidgeh* **43**, 62-68.

Wohlfarth, G.W., Moav, R. & Hulata, G. 1975. Genetic differences between the European and Chinese races of the common carp. II. Multi-character variation - a response to the diverse methods of fish cultivation in Europe and China. *Heredity* **34**, 341-350.

Wohlfarth, G.W., Moav, R. & Hulata, G. 1983. A genotype-environment interaction for growth rate in the common carp growing in intensively manured ponds. *Aquaculture* **33**, 187-195.

Wohlfarth, G.W., Moav, R. & Hulata, G. 1986. Genetic differences between the Chinese and European races of common carp. 5. Differential adaptation to manure and artificial feeds.*Theoretical and Applied Genetics* **72**, 88-97.

Wohlfarth, G.W., Feneis, B., Lukowicz, M.von & Hulata, G. 1984. Application of selective breeding of the common carp to European aquaculture. *European Aquaculture Society Special Publication* **6**,177-193

Wohlfarth, G.W., Lahman, M., Hulata, G. & Moav, R. 1980. The story of Dor-70, a selected strain of the Israeli common carp. *Bamidgeh* **32**, 3-5.

Yu, T.C., Lay, J.Y. & Lin, D.Y. 1987. Study of triploidy induced by cold shocking on *Cirrhina molitorella. Bulletin of Taiwan Fisheries Research Institute* **43**, 165- 169.

Zhongbo, M., Xingzhong, Z., Qianru, Q., Shuhua, L. & Jiansen, Z. 1981. The analysis of the economic characteristics of a hybrid from two ecological types of carps. *Journal of Fisheries of China* **5**, 187-198. (In Chinese with English abstract).

Zobel, H. 1989. Aus der Binnenfischerei der CSSR. *Zeitschrift der Binnenfischerei der DDR* **36**:,183-195.

Zonova, A.S. & Shatnova, Z.A. 1970. Selective breeding of the northern hybrid carp in the Novgorod region. In *Selective Breeding of Carp and Identification of Fish Breeding in Ponds* (Kirpichnikov, V. S. ed.) pp.140-159. Washington D.C.: National Science Foundation.

8

THE NILE TILAPIA

Ambekar E. Eknath

INTRODUCTION

Ideally, management of aquatic genetic resources should involve a continuum of activities: *documentation* of genetic resources and the variety of ecosystems of which they are functioning parts, including status of and potential threats to natural and farm stocks; *characterization* to determine the genetic structure or distinctness and conservation value of the resource; *evaluation* to estimate either direct economic or indirect economic potential; and *utilization* in sustainable breeding schemes, including the politics of access to germplasm. The common theme of these activities should be *conservation* of genetic diversity.

Unlike the other management examples described in this volume, the tilapias (Plate 4) represent a special scenario. The natural tilapia genetic resources are restricted to Africa and the Levant, whereas the main aquaculture industries are at present in Asia. While the African countries hold the global wealth of tilapia genetic resources but contribute only about 5% to the global tilapia production, the tilapia culture industry in Asian and other non-African countries is based on a very narrow genetic base from a few founder populations (Pullin, 1988). Moreover, many natural populations of tilapia in Africa are under severe threat of irreversible change or loss from factors such as fish transfers and habitat disturbance.

The amount of information on tilapias compiled from various field and laboratory studies is vast. The major sources of references and bibliographies are Chimits (1955), Thys van den Audenaerde (1968), Fryer & Iles (1972), Balarin & Hatton (1979), Balarin & Haller (1982), Pullin & Lowe-McConnell (1982), Schoenen (1982, 1984, 1985), Fishelson & Yaron (1983), Wohlfarth & Hulata (1983), Pullin (1988), Pullin *et al.* (1988, and in press). In addition, Lai & Huang (1981) published a bibliography from Taiwan, and a compilation of papers on tilapias in Russian is available from A.Ivoilov at the Biological Institute, University of St. Petersburg. On the basis of a comprehensive analysis of research results from about 3000 documents, Pullin & Maclean (1992) have concluded that there is a glaring lack of interdisciplinary approach to the development of tilapia farming. Nevertheless, these publications contain a wealth of information on the biology of tilapias. Trewavas's (1983) monograph on tilapiine fishes is basic for studies of mechanisms of evolution and for understanding biodiversity.

This review is limited to what is known about Nile tilapia genetic resources along the lines suggested above: documentation, characterization, evaluation, utilisation, and conservation. A pilot study in the Philippines on the Genetic Improvement of Farmed Tilapias (GIFT) being implemented by ICLARM is described here to demonstrate how stepwise progression from documentation of genetic resources through their evaluation to their utilization in breeding programmes can generate rapid benefits for both farmers and consumers. A brief section on the role of tilapias in aquaculture is included to provide development context.

ROLE OF TILAPIAS IN AQUACULTURE

Tilapias possess remarkable attributes for aquaculture (Pullin, 1985): excellent growth rates on low protein diets whether cropping natural aquatic production or receiving supplementary feeds; tolerance of wide ranges of environmental conditions; comparative freedom from serious diseases and parasitic infections; ease of handling and breeding in captivity and wide acceptance as food fish. Tilapias are therefore recognized as prime domesticated species for farming in a wide range of aquaculture systems from simple waste-fed backyard fish ponds to intensive feed-lot systems. They form the mainstay of small-scale aquaculture for many farmers in the developing world. Tilapias have been dubbed the aquatic chicken (Maclean, 1984).

Although several tilapia species are cultured, the most widely preferred (in over 40 countries) is the Nile tilapia *Oreochromis niloticus*. It shows better growth rates and greater feeding versatility than others. It possesses fine gill rakers for filter feeding on plankton, and stout pharyngeal and jaw teeth for herbivory and detritivory and accepting supplemental foods such as cereal brans and pellets (for a review on feeds and feeding, see Jauncey & Ross, 1982).

Tilapias are farmed in the tropics and subtropics of all continents. FAO statistics report tilapia culture in 68 countries (Pullin, in press). Over 95% of total production comes from farming of maternal mouthbrooders (*Oreochromis* spp.), particularly the Nile tilapia. In Asia, tilapia production contributes about 10% of the total aquaculture production of finfish. This is an underestimate because production statistics from several countries, including Bangladesh and Vietnam, are not readily available. There is significant tilapia culture in Bangladesh, China, Indonesia, the Philippines, Thailand, and Vietnam. Interest in tilapia culture is also increasing elsewhere in the Indian subcontinent and in the Pacific-rim countries. In Africa, although the total aquaculture production is low, tilapia production is dominant (about 65% of the total aquaculture production of 8000 metric tons). The same is true for the Latin American/Caribbean region. The constraints to expansion of tilapia culture have been dealt with extensively by Pullin (in press). These include: negative attitudes and policies, poor breeds, poor non-sustainable farming systems, and possible adverse environmental impacts. Interdisciplinary approaches to match development efforts and the needs and circumstances of producers are also lacking (Pullin & Maclean, 1992).

MANAGEMENT OF NILE TILAPIA GENETIC RESOURCES

Documentation

Natural distribution and transfers

Trewavas (1983) has summarized the ecological diversity of Nile tilapia. Its natural distribution is the River Nile and its tributaries and lake systems in the north; the watersheds of the lakes of Tanganyika in the south; the Rivers Niger, Volta, Gambia and Sénégal in the west; the watersheds of sub-Sahelian West Africa; and the Jordan Valley. The geographical distribution of the Nile tilapia extends from 8°S to 32°N, and from sea level to 1830 metres.

However, the original distribution has been greatly modified by deliberate and often unplanned introduction. Philippart & Ruwet (1982) have compiled some of the known examples of introductions of several tilapia species outside their original distribution. Such transfers and introductions had a variety of objectives (Philippart & Ruwet, 1982): stocking natural lakes in which the species concerned did not occur (e.g., Nile tilapia into many lakes in Uganda and Rwanda); introductions to fill an ecological niche (e.g., Lake Victoria and Lake Kyoga); introductions to develop new fisheries (e.g., reservoirs of a southern Tunisian oasis); for biological control of mosquitos; and for development of aquaculture. A number of involuntary introductions have also occurred (during deliberate introductions) mostly owing to confusion between sympatric species.

From the list of introductions (within and outside Africa) and their impacts on various ecosystems compiled by Philippart & Ruwet (1982), it appears that introductions into water bodies not containing any tilapias or offerring a vacant ecological niche have been successful (e.g., Nile tilapia in Uganda (Koki Lakes) and in Madagascar; augmentation of local reservoir fishery resources in Indonesia, Bangladesh and Mexico). However, many introductions especially into waters already containing indigenous tilapias have had catastrophic consequences on aquatic ecology and on the fisheries. An example is Lake Victoria (details in Lowe-McConnell, 1982, and discussions led by Lowe-McConnell in Pullin, 1988). Briefly, in an attempt to boost the tilapia fisheries, *Tilapia zillii* from Lake Albert was introduced to Lake

Victoria in 1954, and along with it came *O. niloticus* and *O. leucostictus*. *Oreochromis niloticus* may have entered the lake also from culture trials in areas that drain into Lake Victoria. The general result has been that *O. niloticus* has displaced almost completely or hybridized with the endemic *O. esculentus*. Also, *T. zillii* has largely displaced the endemic *O. variabilis*. A similar example comes from Lake Itasy, Madagascar, where *O. macrochir* (introduced in 1958) prospered for several years before disappearing and being replaced by *O. niloticus* (introduced in 1961-62). Hybridization of the two species has produced slow-growing and sometimes deformed individuals (Lamarque *et al.*, 1975, cited in Philippart & Ruwet, 1982).

Furthermore, a number of examples summarized by Lowe-McConnell (1982) show that the same species introduced into different waters often respond in an unpredictable manner depending on the prevailing environmental characteristics: fish from large lakes grow to and mature at larger sizes than those from lagoons and ponds; in larger lakes there is no significant sexual dimorphism in growth and maturation size, but in small water bodies males are larger than females; whereas fish from large lakes are relatively free of parasites, the fish from shallow waters (usually dwarfed populations) are often heavily parasitized. Fishing pressure has also brought about significant changes in the biology of the fish. For example, in Lake George, Uganda, fishing pressure reduced the size at maturation of *O. niloticus* (*O. n. eduardianus*) over an eleven year period (Gwahaba, 1973, cited by Lowe-McConnell, 1982). Early in the history of the fishery, 50% of the females were mature at 27.5 centimetres, by 1960 this had fallen to 24.5 centimetres, and in 1972 was only 20 centimetres.

Clearly most of the introductions and transfers have been made without sufficient knowledge of the biology of the species and the ecosystems. A databank is needed urgently to document the precise geographical distribution of populations and the introductions and transfers made so far. This will enable the effects of such transfers to be evaluated and provide a reference point for understanding the evolution of the populations concerned. It should be recognized that the natural populations in Africa are the *in situ* repository of genetic diversity, which needs to be maintained. However, introductions and transfers will be essential for the future development of the aquaculture industry. In all such future planned transfers and introductions competent technical bodies should be involved to advise on and control introductions and transfers.

Taxonomy

As in other groups of fishes, taxonomy in tilapias is not without its problems. The recognition of the level of differentiation necessary to assign a given population to a new species, subspecies or strain has been a matter of long-standing debate - the classic case of lumpers versus splitters (Thys van den Audenaerde, in Pullin, 1988). Such debates are inevitable especially when water bodies become isolated and the populations therein change through natural selection, mutation, and genetic drift. This situation is complicated further by indiscriminate fish transfers, interbreeding between wild and cultured fish populations, and inter- and intraspecific hybridization programmes.

According to Thys van den Audenaerde, the nomenclature that applies to natural populations is of little use for labelling cultured tilapias (in Pullin, 1988). For natural populations the nomenclature proposed by Trewavas (1983) is followed widely. There are now seven recognised subspecies of *Oreochromis niloticus: O. n. niloticus* (inhabiting the River Nile system); *O. n. eduardianus* and *O. n. cancellatus* (common in the East African lakes and the Ethiopian lakes, respectively); *O. n. vulcani* (crater lakes of Lake Turkana and its environs); *O. n. filoa* (hot alkaline springs in the Awash system); *O.n.baringoensis* (Lake Baringo, Kenya); and *O. n. sugutae* (Suguta River and the surrounding areas).

The number of subspecies not only suggests that *O. niloticus* is an 'evolving supraspecies' (Pullin, 1988), but it is testament to its inherently high diversity, and one way that this is expressed is in its extraordinary physiological tolerances. The wide range of habitats where this species occurs and has been introduced represents, both in terms of absolute amplitude and in terms of the speed at which fluctuations take place, a varied range of physical and chemical parameters. Balarin & Hatton (1979) and Philippart & Ruwet (1982) have compiled the extensive literature concerning tolerance limits and preferences of tilapias. For example, *O. niloticus* is capable of maintaining populations in habitats where the salinity reaches as much as 30‰ (in Bitter Lakes and Lake Qarun, Egypt); where pH ranges from 8 to 11 (in Sudanese ponds); and where dissolved oxygen levels are as low as 0.1 parts per million - a characteristic that allows them to grow and reproduce in shallow lakes and swampy areas. *O. niloticus* inhabiting waterways, irrigation canals, rice paddies, water holes on farms, etc., does not seem to be affected by the high levels of chemicals that are usually found in such habitats. However, colonization in some of these unpromising waterways may have brought about, over a period of several generations, a variety of responses in terms of growth, maturation and morphological characteristics. A systematic study of such populations should provide insight into the speciation.

The plasticity of growth and maturation of *O. niloticus* within its natural distribution range has been reviewed by Lowe-McConnell (1982, and in Pullin, 1988). For example, in Lake Albert, which is connected with the Nile River, *O. n. eduardianus* grows to very large sizes, but populations trapped in lagoons are of very small fish. In such populations the maturation size is also very small, and the females are much smaller than the males. All such fish have very poor condition factors compared with fish from the main lake. However, in the main lake the males and females grow to, and mature at, about the same size. Therefore the same 'strain' grows to a large size in the lake but to a much smaller size in the lagoons. One important conclusion from a number of such observations is that the natural distribution of a given population may not be a reliable guide to the environmental tolerance of that population or to its growth or maturation performance outside its natural distribution. Clearly, systematic research on the relative importance of genetics and environmental factors and their interactions is needed. Moreau *et al.*, (1986) have derived an index of growth performance that allows the growth potential of different populations to be compared.

Aquaculture stocks

Nearly all domesticated aquaculture stocks supporting the expanding tilapia culture industry have been derived from very small founder populations. For example, the early transfers (1956-68) into francophone Africa (Bouaké, Côte d'Ivoire) were all derived from a small isolated population inhabiting a waterhole in Burkina Faso (Thys van den Audenaerde, in Pullin, 1988). From Bouaké, transfers have been made to many other destinations. The shipment to Brazil in 1971 contained only 60 juveniles from a founder population of small size (Nugent, in Pullin, 1988). Subsequently, Auburn University received a founder population of about 100 fry from Brazil (Smitherman, in Pullin, 1988). A founder stock collected from the wild in Egypt in 1962 was transferred to Japan and its descendants used for transfer to Thailand in 1965, and from there to the Philippines in 1972. This 'strain' is still used by tilapia farmers, known as the 'Chitralada' strain in Thailand, and 'Thailand' strain in the Philippines (Pullin, in press).

Pullin & Capili (1988) have attempted to trace the history of introductions and subsequent transfers of Nile tilapia in Asia. One important conclusion of this study is that the aquaculture stocks in Asia are descendants of a few introductions, mostly through intermediate non-tropical sources, consisting of very few fish, and are probably suffering from genetic bottleneck effects. For example, after nearly 26 years (or about 50 generations) under domestication the performance of the Thailand strain in the Philippines, across a wide range of environments, is inferior to or similar to that of a new founder stock of wild fish from Egypt collected in 1988 (Eknath *et al.*, 1993). Genetic deterioration is widespread owing to generations of poor broodstock management resulting in inbreeding and introgression of genes from other less desirable feral tilapia species (mainly *O. mossambicus*, the original ambassador to Asia). Furthermore, a number of hybrids of different attributes (colours, shapes) are also produced.

While the tilapia genetic resources in Africa are threatened by indiscriminate fish transfers, the situation elsewhere poses a different set of vexing problems for researchers. There is a growing tendency among aquaculturists and breeders to label their strain with the names of institutions, companies, or even individuals. This has frustrated researchers who are already confronted with poorly characterized materials of largely unknown origins. Clearly a standardized nomenclature is essential. The participants at the Workshop on Tilapia Genetic Resources for Aquaculture, Bangkok 1987 (discussion and recommendations in Pullin, 1988) have suggested strongly that strain registries should be established along the lines of cereal crops. A pioneering attempt in this direction is the FishBase project in ICLARM (Augustin *et al*., 1993).

Strain Registry and FishBase

FishBase is a large relational database under development at ICLARM in cooperation with FAO and with support from the Commission of the European Community (CEC) (Froese, 1990). Among its many features is the global information on fish genetic resources, including information on nomenclature, physiology, population dynamics, diseases, and conservation status. FishBase will eventually contain an inventory of all recorded collections of Nile tilapia throughout its native range and information on all international introductions and transfers. It will provide genetic data, including the composition of founding stocks and their subsequent management, accumulated levels of inbreeding, and genetic characterization of strains and hybrids (Augustin *et al*., 1993).

One important problem that has to be dealt with during the setting up of a strain registry is the definition of what constitutes a strain. There is no unequivocal definition of a strain. Strain registry will also be fraught with difficulties owing to species introgressions. Furthermore, the strain itself will be under constant genetic change because of the intervention of fish breeders, natural selection, and genetic drift. Therefore the data on a given strain may become obsolete as soon as they are entered in the registry. Nevertheless, strain registries will provide information of great value to fish breeders and policy makers concerned with conservation issues.

To standardize strain nomenclature, FishBase uses a unique code composed of 7 letters and a 3-digit number (Augustin *et al*., 1993). An example is ORNILNI001 for a strain of *O. n. niloticus*. The first two letters refer to the first two letters of the genus, letters 3-5 refer to the first three letters of the species, and letters 6-7 refer to the first two letters of the subspecies. The 3-digit number is sequential. For unknown subspecies, letters 6-7 will be designated as XX and for hybrids, letters 6-7 will be HX. Initial efforts are focused on summarizing and entering data on protein polymorphism from electrophoretic studies (including levels of heterozygosity and proportions of polymorphic loci), using the nomenclature recommended by the International Union of Biochemistry's Nomenclature Committee (Shaklee *et al.*, 1990).

Characterization

To address the taxonomic problems and also to identify the distinctness of both natural and aquaculture populations, a variety of techniques has been used: morphometrics (Pante *et al.*, 1988; Velasco *et al.*, in press); electrophoresis (McAndrew & Majumdar, 1983; Taniguchi *et al.*, 1985; Macaranas *et al.*, 1986; Galman *et al.*, 1988; and many others); serum protein analysis (Avtalion *et al.*, 1976); immunology and agglutination assays (Avtalion & Wojdami, 1971; Oberst *et al.*, 1989); mitochondrial DNA restriction analysis (Seyoum, 1989); karyotype analysis (Crosetti *et al.*, 1988); and DNA fingerprinting (Harris *et al.*, 1991). A comprehensive description of some of these techniques and their applications is given by Kornfield (1991).

Briefly, multivariate analyses of morphometric (using truss network of landmark points on the body outline) and meristic characters can discriminate species, but not between different strains within a species (Pante *et al.*, 1988; Velasco *et al.*, in press). The practical utility of electrophoresis as a technique to distinguish natural populations has been demonstrated by McAndrew & Majumdar (1983). The widespread introgression of *O. mossambicus* genes into commercial stocks of *O. niloticus* in the Philippines was confirmed by electrophoresis (Taniguchi *et al.*, 1985; Macaranas *et al.*, 1986). The Philippine domesticated populations of Nile tilapia were genetically more diverse than the wild Egyptian ones. This is probably due to the recombinations that are characteristic of introgressed populations. Serum protein analyses have shown sex differences (Avtalion *et al.*, 1976) only in the subpecies *O. n. vulcani* but not in others. Mitochondrial DNA characterisations could distinguish unambiguously the seven subspecies of *O. niloticus* (Seyoum, 1990). Immunological techniques and blood group studies have been successful also in discriminating species (Oberst *et al.*, 1992) and a field kit is now under preparation through collaborative research between the Institute of Aquatic Biology (Ghana), the Zoologisches Institut und Museum, Universität Hamburg (Germany) and ICLARM.

One of the advantages of DNA sequencing over other methods is data standardization. Although only a few results are available so far, the potential is vast. Some of the applications include identification of genetic lineages, estimation of inbreeding rates in aquaculture stocks, monitoring fish transfers, and securing breeders' rights.

Clearly, the methodology for characterization of tilapia genetic resources is readily available. Through analysis of genetic distances and cladistic relationships, the identity of natural and aquacultural populations has been established. However, methods to assign conservation value to a given genetic resource are needed. This is a universal problem that should be tackled globally. Typically, methods are needed to determine whether and when a particular population or genetic resource base should be conserved. In the short term, and strictly from a breeder's point of view, screening for and estimating the extent of genetic variation and covariation between traits of economic importance, genotype x environment interactions, and identifying genetic lineages will be of great benefit for establishing sound breeding schemes.

A topic that deserves special mention here is that of the sex determination mechanism in tilapias. In aquaculture, male tilapias grow 20-40% larger than females. Therefore in almost all progressive tilapia culture systems the goal is to grow an all-male population in order to prevent early maturation and breeding (a major negative attribute of tilapias in confined conditions) and to capitalize on the faster growth of males over females. The various methods of population control in farmed tilapias have been reviewed by Mair & Little (1991). The methods include: sex reversal by androgenic hormones, intermittent harvesting, manual sexing, use of predators to control recruitment, high density stocking, cage culture, delayed sexual maturity, sterilization, hybridization, and the production of YY-broodstock. The most widely used methods include hormonal sex reversal, hybridization, and a combination of both.

One of the research areas that most tilapia biologists are engaged in is the mechanism of sex determination (Trombka & Avtalion, 1993). Briefly, four different approaches have been used: interspecific and intraspecific crosses; sex inversion of fry to females or males by hormonal treatment; chromosomal manipulations leading to ploidy and gynogenesis; and karyotyping and differential staining of the tilapia genome. On the basis of evidence gathered, several models of sex determination have been proposed (Wohlfarth & Wedekind, 1991). A major hurdle in studies on sex determination mechanisms is the lack of distinct sex-linked markers in tilapias. The general theme of most of the models is that, principally, sex determination follows a distinct gonosome strategy, albeit with autosomal and environmental influences. However, Wohlfarth & Wedekind (1991) suggested that sex ratio be treated as a quantitative trait for sex determination studies.

Evaluation

Evaluation of tilapia genetic resources has been restricted so far to assessing their aquaculture potential, in particular their growth performance in a range of aquaculture environments (Wohlfarth & Hulata, 1983; Fishelson & Yaron, 1983; Pullin *et al.*, 1988; Pullin *et al.*, in press). Refinements in methodology for comparative evaluation of genetic resources are becoming available (discussion by Wohlfarth and by Smitherman in Pullin, 1988; Eknath *et al.*, 1993; Palada-de Vera & Eknath, 1993). Nile tilapia stocks have been screened for important traits such as growth, tolerance to cold temperatures and a range of salinities, and production of all-male hybrids with *O. aureus*. The limited studies made on estimation of genetic parameters indicate considerable genetic variation within strains for important traits such as growth. Where careful experimental designs have been used, a good response to selection has been demonstrated (Plate 4, and see GIFT project, below). For further systematic evaluation, definition and recording procedures for other traits of economic importance such as maturity, resistance to disease and parasitic infestations are needed. Perhaps a combination of techniques that are used in screening other finfish species could be adapted for Nile tilapia.

However, a major problem is that the majority of evaluation experiments are based on stocks (usually of unknown or doubtful origin) derived from the introduction of small numbers of fish. The results so obtained may not have any relevance to the natural tilapia genetic resources. A systematic evaluation of the tilapia genetic resources from their natural distribution is lacking. Screening of natural tilapia genetic resources is clearly beyond the scope of any single research group or institution. International cooperation is necessary to organize a well coordinated evaluation programme. The recently established International Network on Genetics in Aquaculture (see below), coordinated by ICLARM, is a step in this direction.

Utilization

The Nile tilapia is poised to become an international commodity (Davlin, 1991). The use of Nile tilapia as an important food fish is growing in a number of developing countries (Pullin, in press). Pullin (1991) forecast a doubling of world tilapia production over the next 10 years. To realize this potential, however, research should be directed towards farmers' needs, changing the anti-tilapia attitude, and most importantly tilapia farming should become a more sustainable and environmentally compatible enterprise, well integrated with other development initiatives (Pullin, in press).

As discussed above, there is a need to promote interregional cooperation between African countries that hold the global wealth of Nile tilapia genetic resources and other non-African countries where Nile tilapia is exotic but forms an important component of the aquaculture industry. At present, the aquaculture performance of farmed stocks of Nile tilapia in Asia is close to wild types or worse. There is a need to transfer germplasm from Africa to develop improved breeds.

In the light of the United Nations Conference on Environment and Development (UNCED) and the founding of the Biodiversity

Convention, the relevant issues are: potential socioeconomic and environmental impacts of international transfer of germplasm and improved breeds, access to germplasm, and how to compensate Africa for the use of its genetic resources. The challenge is to develop methods taking into consideration the lessons learned from the Green Revolution. These issues formed the background for a meeting on 'International Concerns in the Use of Aquatic Germplasm' organised by ICLARM under the auspices of the GIFT project in June 1992.

To address the issue of environmental effects of introducing germplasm and improved breeds, the international panel of experts recommended that ICLARM promote establishment of databases on fish ecology with a focus on resident biota and trophic interactions between fishes. ICLARM's FishBase is already developing this. The other important question regarding distribution of germplasm is whether it may be better to export methodology rather than improved breeds. ICLARM actively encourages client countries to develop their own breeding programmes to suit particular local needs and to use the experience of the GIFT project as a possible model. National control of germplasm dissemination and the involvement of non-governmental and farmers' organizations seem to be the appropriate mechanisms to reach small scale farmers. At present one of the mechanisms to compensate African countries for their contribution to the development of tilapia aquaculture is to increase the commitment to African aquacultural development.

Conservation

Although the practical utility of conservation of aquatic genetic diversity is widely accepted, translation of ideas into a plan of action has been difficult. For tilapias, both *in situ* and *ex situ* conservation measures are possible. There should be complementarity between long-term *in situ* conservation and *ex situ* conservation. Pullin (1990) has elaborated the status and issues involved in conserving tilapias as follows.

In situ *conservation*

The best strategy for conservation is to maintain the original habitats (Pullin, 1990). The first priority is better documentation of the status of the genetic diversity and greater awareness of its value. Documentation on the tilapias and their habitats has just begun mostly through databases like the FishBase, which already incorporates a tilapia strain registry. An essential prerequisite to conservation as a responsible attitude to fish transfers with due regard to quarantine and possible environmental impacts (Pullin, 1988). National commitment to the cause of conservation is of paramount importance. However, the obstacles are great. As yet, aquaculture has little economic importance in Africa. Many of the large water bodies are shared between nations. Malawi and Ghana are examples of nations with responsible attitudes and important fish populations. Malawi is making all possible efforts to conserve the Lake Malawi ecosystem and Ghana has established a nature reserve on an ecologically important sector of the Volta catchment (Pullin, 1990).

Ex situ *conservation*

At present only live fish and sperm banks are possible. A number of institutions are maintaining small live fish collections and sperm banks mostly for research purposes. ICLARM has initiated a live fish and sperm bank for the African strains that it collected for developing a national breeding programme in the Philippines. ICLARM's strategy is to collect *O. niloticus* subspecies from different river basins. Although *ex situ* conservation will have an important role to play in research and development of breeding programmes, it should be recognised the collections are costly to establish and maintain. Nevertheless a code of practice is essential for obtaining and maintaining live fish collections.

THE GIFT PROJECT - A PILOT STUDY

In the mid-1980s, ICLARM started to explore a research programme to develop improved breeds of fish grown in inland aquaculture systems by small scale farmers. Tilapias were chosen as test species because of their growing importance in warm water aquaculture in Asia and Africa, and their utility in investigating the application of genetics in aquaculture, from conservation of genetic resources to breeding programmes.

The GIFT project was established in 1988 through collaboration among the National Freshwater Fisheries Technology Research Center of the Philippine Bureau of Fisheries & Aquatic Resources (NFFTRC/BFAR), the Freshwater Aquaculture Center of the Central Luzon State University (FAC/CLSU), the Marine Science Institute of the University of the Philippines (UPMSI), AKVAFORSK of Norway and ICLARM. Thus the expertise to implement the project was secured by bringing together institutions with experience in: tilapia farming (BFAR, FAC); the genetic status of wild and cultured tilapia stocks (UPMSI, ICLARM); and application of genetic improvement strategies (AKVAFORSK, ICLARM). Support came from the Asian Development Bank and the United Nations Development Programme/Division for Global and Interregional Programmes (UNDP/DGIP). The project plans were circulated for peer review to 25 international experts before implementation. The project's primary objective is to develop effective approaches and methods for the production of improved breeds of Nile tilapia for low-cost sustainable aquaculture.

The Process - From systematic documentation to improved breeds

Because of the pioneering role of the GIFT project, an elaborate design was chosen to seek answers to a broad range of basic problems in fish breeding programmes. The team had to invent and adapt methods for controlled mating, tagging, frequency of sampling, recording of traits, managing the databases, and rigorous statistical analyses. Despite setbacks from a series of natural calamities in the Philippines, the GIFT team followed an exacting sequential approach including: systematic documentation of the poor status of Asian cultured stocks, collection and transportation of new germplasm from Africa to Asia with thorough quarantine; rigorous evaluation of promising strains; establishment of a base population; and plans to develop improved breeds of tilapia for a wide range of farming systems and agroclimatic conditions.

The background work for the project was initiated during a special Workshop on Tilapia Genetic Resources for Aquaculture in 1987 (Pullin, 1988) which concluded that the established farm stocks in Asia do not form the best genetic base for a genetic improvement programme because of their small founder populations, possible inbreeding and introgression of genes from other tilapias (Macaranas *et al.*, 1986; Pullin & Capili, 1988). These problems were documented and decisions were taken to collect new germplasm from Africa: Egypt, Ghana, Kenya and Sénégal. The strategy was to evaluate new germplasm along with four established farmed stocks in the Philippines in diverse tilapia farming systems.

From each country, 150-160 breeders or 200-800 fingerlings were shipped. On arrival the fish were held for 3 to 7 months in quarantine.

The project's first generation experiment was focused on estimating the magnitude of genotype x environment (GE) interaction. In this experiment, 11,400 tagged fingerlings were reared communally for 90 days in 11 different test environments, representing a wide range of tilapia farming systems. The GE interaction was low, indicating no need to develop specialized strains for each of the different farming systems used in the test. Some of the African wild strains grew faster than the farmed strains (Eknath *et al.*, 1993). This was followed by a complete 8x8 diallel crossing experiment, producing all the possible hybrid crosses among the strains in order to estimate the magnitude of heterosis or hybrid vigour, and thereby the breeding strategy: selection or cross-breeding. About 21,000 tagged fingerlings were tested communally in eight environments. The gain in growth and survival by cross-breeding was too low to be of significance in an applied breeding programme.

A simple selection strategy was then started by selecting the best growing breeders from the 25 best performing pure-bred and cross-bred groups (out of the 64 evaluated) to build a genetically mixed base population. The synthetic breed served as the base for further generations of selection. The individuals most likely to produce the best offspring were chosen then as breeders in each successive generation. A combined family and within-family selection strategy was adopted. The potential breeders were ranked based on breeding value (the additive genetic value of an individual), judged by performance of the individual itself and its full- and half-sibs. From the third generation onwards, about 20,000 tagged fingerlings from 120-200 full-sib families (within 50-100 half-sib families) have been tested communally in each generation in a variety of test environments.

After one generation of selection in the synthetic breed, the selected fish grew 23% faster than the previous generation and 75% faster than the most common commercial strain in on-station trials (Plate 4). In on-farm trials, the GIFT fish grew on average 60% faster than the farm breeds. Their survival was almost 50% better. A survey in three provinces in the Philippines indicated that presently farmed breeds reach a harvest size of about 80 grammes after a growing period of 5-6 months, permitting about two crops per year. The GIFT fish reach the same harvest size in 3 months, permitting three crops per year. This can translate into increases in annual production and gross revenue. Furthermore, the improved fish to be disseminated to farmers will include, through selection, a cumulative increase in growth rate of about 10% per generation (about 8 months).

Cognisant of the positive and negative lessons learned from plant and livestock breeding and the need to seek advice from international experts on the potential socioeconomic and environmental impacts of improved breeds, the GIFT project convened a meeting on International Concerns in the Use of Aquatic Germplasm. The recommendations of this meeting (see utilization section above) led to the development of an Integrated Breeding and Dissemination Plan that takes into account conservation of biodiversity and indigenous knowledge on genetic resources.

National breeding programmes

So far there are almost no sustained applied fish breeding programmes in developing tropical countries. The GIFT project is therefore a pioneering effort. It has stimulated the Philippines to develop a self-sustaining national breeding programme.

Genetic gains will be disseminated through national broodstock centres (breeding nuclei) and nationwide satellite stations (multipliers), with extensive participation of farmers. The project pictures a farmer-run tilapia breeding programme in the Philippines after a few years and the initiation of similar efforts in other countries.

International networking

The achievements of the GIFT project have encouraged its principal donors (UNDP/DGIP) to assist ICLARM in establishing an International Network on Genetics in Aquaculture (INGA), modelled after a similar network for rice breeding. INGA was established formally in July 1993, with 11 countries collaborating initially in research activities: Bangladesh, China, Côte d'Ivoire, Egypt, Ghana, India, Indonesia, Malawi, the Philippines, Thailand and Vietnam. Collaboration will be sought with other countries and advanced scientific institutions in developed and developing countries.

REFERENCES

Augustin, L.Q., Froese, R., Eknath, A.E. & Pullin, R.S.V. 1993. Documentation of genetic resources for aquaculture - the role of FishBase. In *International Workshop on Genetics in Aquaculture and Fisheries Management* (Penman, D., Roongratri, N. & McAndrew, B. eds.) pp.63-68. Bangkok, Thailand: ASEAN-EEC Aquaculture Development and Coordination Programme.

Avtalion, R.R. & Wojdami, A. 1971. Electrophoresis and immunoelectrophoresis of sera from some known Fl hybrids of tilapia. *Bamidgeh* **23**, 117-124.

Avtalion, R.R., Duczyminer, M., Woydjani, A. & Pruginin, Y. 1976. Determination of allogenic and xenogeneic markers in the genus Tilapia. II. Identification of *T. aurea*, *T. vulcani* and *T. nilotica* by electrophoretic analysis of their serum proteins. *Aquaculture* **7**, 255-265.

Balarin, J.D. & Haller, R.D. 1982. The intensive culture of tilapia in tanks, raceways and cages. In *Recent Advances in Aquaculture* (Muir, J.F. & Roberts, R.J. eds.) pp.266-355. London: Croom Helm.

Balarin, J.D. & Hatton, J.P. 1979. *Tilapia: A guide to their biology and culture in Africa.* Stirling, Scotland: University of Stirling.

Chimits, P. 1955. Le tilapia et son élevage: bibliographie préliminaire. *FAO Fisheries Bulletin* **8**, 1-38.

Crosetti, D., Sola, L., Brunner, P. & Cataudella, S. 1988. Cytogenetical characterization of *Oreochromis niloticus*, *O. mossambicus* and their hybrid. In *The Second International Symposium on Tilapia in Aquaculture* (Pullin, R.S.V., Bhukaswan, T., Tonguthai, K. & MacLean, J.L. eds.) Manila, Philippines: International Center for Living Aquatic Resources Management, Conference Proceedings **15**,

Davlin, A.Jr. 1991. The nineties: a booming decade for the aquaculture industry! *The Aquaculture Industry, an analyst's report.* Vol.III (1), 1-6.

Eknath, A.E., Tayamen, M.M., Palada-de Vera, M.S., Danting, J.C., Reyes, R.A., Dionisio, E.E., Capili, J.B., Bolivar, H.L., Circa, A.V., Bentsen, H.B., Gjerde, B., Gjedrem, T. & Pullin, R.S.V. 1993. Genetic improvement of farmed tilapias: the growth performance of eight strains of *Oreochromis niloticus* tested in different farm environments. *Aquaculture* **111**, 171-188.

Fishelson, L. & Yaron, Z. (eds.) 1983. *Proceedings of the First International Symposium on Tilapia in Aquaculture*. Tel Aviv, Israel: University of Tel Aviv.

Froese, R. 1990. FishBase: an information system to support fisheries and aquacultural research. *Fishbyte* **8**(3), 21-24.

Fryer, G. & Iles, T.D. 1972. *The Cichlid Fishes of the Great Lakes of Africa: their Biology and Evolution.* Edinburgh: Oliver and Boyd.

Galman, O.R., Moreau, J., Hulata, G. & Avtalion, R.R. 1988. The use of electrophoresis as a technique for the identification and control of tilapia breeding stocks in Israel. In *The Second International Symposium on Tilapia in Aquaculture* (Pullin, R.S.V., Bhukaswan, T., Tonguthai, K. & MacLean, J.L. eds.) pp. 177-182. Manila, Philippines: International Center for Living Aquatic Resources Management, Conference Proceedings **15.**

Harris, A.S., Bieger, S., Doyle, R.W. & Wright, J.M. 1991. DNA fingerprinting of tilapia *Oreochromis niloticus*, and its application to aquaculture genetics. *Aquaculture* **92**, 157-163.

Jauncey, K. & Ross, B. 1982. *A Guide to Tilapia Feeds and Feeding*. Stirling, Scotland: University of Stirling.

Kornfield, I. 1991. Genetics. In *Cichlid Fishes: Behaviour, Ecology and Evolution* (Keenleyside, M.H.A. ed.) pp.103-108. London: Chapman & Hall.

Lai, C.F. & Huang, L.C. 1981. A bibliography of *Tilapia* (family Cichlidae) in Taiwan. *Aquaculture* **22**, 389-394.

Lowe-McConnell, R.H. 1982. Tilapias in fish communities. In *The Biology and Culture of Tilapias* (Pullin, R.S.V. & Lowe-McConnell, R.H. eds.) pp. 83-113. Manila, Philippines: International Center for Living Aquatic Resources Management, Conference Proceedings **7**.

Macaranas, J. M., Taniguchi, N., Pante, M.J.R., Capili, J.B. & Pullin, R.S.V. 1986. Electrophoretic evidence for extensive hybrid gene introgression into commercial *Oreochromis niloticus* (L.) stocks in the Philippines. *Aquaculture & Fisheries Management* **17**, 249-258.

Maclean, J.L. 1984. Tilapia - the aquatic chicken. *ICLARM Newsletter* **7(1)**, 17.

McAndrew, B.J. & Majumdar, K.C. 1983. Tilapia stock identification using electrophoretic markers. *Aquaculture* **30**, 249-261.

Mair, G.C. & Little, D.C. 1991. Population control in farmed tilapias. *Naga ICLARM Quarterly* **14** (3), 8-13.

Moreau, J., Bambino, C. & Pauly, D. 1986. Indices of overall growth performance of 100 tilapia (Cichlidae) populations. In *The First Asian Fisheries Forum* (Maclean, J.L., Dizon, L.B. & Hastillos, L.V. eds.) pp. 201-206. Manila, Philippines: Asian Fisheries Society.

Oberst, S., Villwock, W. & Renwrantz, L. 1989. Antisera from *Tilapia* species to differentiate among erythrocytes from *T. aurea, T. galilaea,* and *T. nilotica* by agglutination assays, and a comparative analysis of hemoglobins. *Journal of Applied Ichthyology* **5**, 18-27

Oberst, S., Villwock, W. & Renwrantz, L. 1992. WESTERN blot analysis of plasma components of the three *Tilapia* species *T. aurea, T. nilotica,* and *T. galilaea*. *Journal of Applied Ichthyology* **8**, 278-292.

Palada-de Vera, M.S. & Eknath, A.E. 1993. Predictability of individual growth rates in tilapia. *Aquaculture* **111**, 147-158.

Pante, M.J.R., Lester, L.J. & Pullin, R.S.V. 1988. A preliminary study on the use of canonical discrimination analysis of morphometric and meristic characters to identify cultured tilapias. In *The Second International Symposium on Tilapia in Aquaculture*, (Pullin, R.S.V., Bhukaswan, T., Tonguthai, K. & Maclean, J.L. eds.) pp. 251-259. Manila, Philippines: International Center for Living Aquatic Resources Management, Proceedings **15**.

Philippart, J.-C. & Ruwet, J.-C. 1982. Ecology and distribution of tilapias. In *The Biology and Culture of Tilapias* (Pullin, R.S.V. & Lowe-McConnell, R.H. eds.) pp. 15-59. Manila, Philippines: International Center for Living Aquatic Resources Management, Conference Proceedings **7**.

Pullin, R.S.V. 1985. Tilapias: everyman's fish. *Biologist* **32** (2), 84-88.

Pullin, R.S.V.(ed.) 1988. *Tilapia Genetic Resources for Aquaculture*. Manila, Philippines: International Center for Living Aquatic Resources Management, Conference Proceedings **16**.

Pullin, R.S.V. 1990. Down-to-earth thoughts on conserving aquatic genetic diversity. *Naga ICLARM Quarterly* **14** (2), 3-6.

Pullin, R.S.V. In press. World tilapia culture and its future prospects. In *The Third International Symposium on Tilapias in Aquaculture* (Pullin, R.S.V., Lazard, J., Legendre, M. & Amon Kathias, J.B. eds.). Manila, Philippines: International Center for Living Aquatic Resources Management, Conference Proceedings **41**.

Pullin, R.S.V. & Capili, J.B. 1988. Genetic improvement of tilapias: problems and prospects. In *The Second International Symposium on Tilapia in Aquaculture* (Pullin, R.S.V., Bhukaswan, T., Tonguthai, K. & Maclean, J.L. eds.) pp. 259-266. Manila, Philippines: International Center for Living Aquatic Resources Management, Conference Proceedings **15**.

Pullin, R.S.V. & Lowe-McConnell, R.H. 1982. *The Biology and Culture of Tilapias*. ICLARM. Manila, Philippines: International Center for Living Aquatic Resources Management, Conference Proceedings **7**.

Pullin, R.S.V. & Maclean, J.L. 1992. Analysis of research for the development of tilapia farming: an interdisciplinary approach is lacking. *Netherlands Journal of Zoology* **42**, 512-522.

Pullin, R.S.V., Bhukaswan, T., Tonguthai, K. & Maclean, J.L.(eds.) 1988. *The Second International Symposium on Tilapia in Aquaculture*. Manila, Philippines: International Center for Living Aquatic Resources Management, Conference Proceedings **15.**

Pullin, R.S.V., Lazard, J., Legendre, M. & Amon Kathias, J.B.(eds.). In press. *The Third International Symposium on Tilapia in Aquaculture*. Manila, Philippines: International Centre for Living Aquatic Resources Management, Conference Proceedings **41**.

Schoenen, P. 1982. *A bibliography of important tilapias (Pisces: Cichlidae) for aquaculture*. Manila, Philippines: International Centre for Living Aquatic Resources Management, Bibliographies **3**.

Schoenen, P. 1984. *A bibliography of important tilapias (Pisces: Cichlidae) for aquaculture*: Oreochromis macrochir, O.aureus, O.hornorum, O.mossambicus, O.niloticus, Sarotherodon galilaeus, Tilapia rendalli *and* T.zillii. Manila, Philippines: International Center for Living Aquatic Resources Management, Bibliographies **3**, Supplement 1.

Schoenen, P. 1985. *A bibliography of important tilapias (Pisces: Cichlidae) for aquaculture:* Oreochromis variabilis, O.andersonii, O.esculentus, O.leucostictus, O.mortimeri, O.spilurus niger, Sarotherodon melanotheron *and* Tilapia sparrmanaii. Manila, Philippines: International Center for Living Aquatic Resources Management, Bibliographies **6**.

Seyoum, S. 1989. Stock identification and the evolutionary relationship of the genera *Oreochromis, Sarotherodon* and *Tilapia* (Pisces: Cichlidae) using allozyme analysis and restriction endonuclease analysis of mitochondrial DNA. Ph.D.Thesis, University of Waterloo, Ontario. Canada.

Shaklee, J.B., Allendorf, F.W., Morizot, D.C. & Whitt, G.S. 1990. Gene nomenclature for protein coding loci in fish. *Transactions of the American Fisheries Society* **119**, 2-15.

Taniguchi, N., Macaranas, J.M. & Pullin, R.S.V. 1985. Introgressive hybridization in cultured tilapia stocks in the Philippines. *Bulletin of the Japanese Society of Scientific Fisheries* **52**, 1219-1224.

Thys van den Audenaerde, D.F.E. 1968. An annotated bibliography of Tilapia (Pisces, Cichlidae). *Documentation Zoologique* **14**. Musée Royale de l'Afrique Centrale, Tervuren. Belgium.

Trewavas, E. 1983. Tilapiine fishes of the genera *Sarotherodon*, *Oreochromis* and *Danakilia*. London: British Museum (Natural History).

Trombka, D. & Avtalion, R.R. 1993. Sex determination in tilapia - a review. *Bamidgeh* **45**, 26-37.

Velasco, R.M.R., Pante, M.J.R., Macaranas, J.N. & Eknath, A.E. In press. Morphometric characterisation of eight Philippine and African *Oreochromis niloticus* strains. In *The Third International Symposium on Tilapias in Aquaculture* (Pullin, R.S.V., Lazard, J., Legendre, M. & Amon Kathias, J.B. eds.) Manila, Philippines: International Center for Living Aquatic Resources Management, Conference Proceedings **41**.

Wohlfarth, G.W. & Hulata, G. 1983. *Applied Genetics of Tilapias.* Manila, Philippines: International Center for Living Aquatic Resources Management, Studies and Reviews **6.**

Wohlfarth, G.W. & Wedekind, H. 1991. The heredity of sex determination in Tilapias. *Aquaculture* **92**, 143-156.

INDEX

Tables in **bold**, Figures in *italic*.